THE BUSINESS OF WAR

For David

The Business of War
Workers, Warriors and Hostages in Occupied Iraq

JAMES TYNER
Kent State University, USA

Routledge
Taylor & Francis Group

LONDON AND NEW YORK

First published 2006 by Ashgate Publishing

Reissued 2018 by Routledge
2 Park Square, Milton Park, Abingdon, Oxon, OX14 4RN
711 Third Avenue, New York, NY 10017, USA

Routledge is an imprint of the Taylor & Francis Group, an informa business

First issued in paperback 2018

A Library of Congress record exists under LC control number: 2006017157

Notice:
Product or corporate names may be trademarks or registered trademarks, and are used only for identification and explanation without intent to infringe.

Publisher's Note
The publisher has gone to great lengths to ensure the quality of this reprint but points out that some imperfections in the original copies may be apparent.

Disclaimer
The publisher has made every effort to trace copyright holders and welcomes correspondence from those they have been unable to contact.

ISBN 13: 978-0-815-39753-3 (hbk)
ISBN 13: 978-1-138-62085-8 (pbk)
ISBN 13: 978-1-351-14756-9 (ebk)

Contents

Contents

Acknowledgements

Having children affects one's sense of purpose. There is, of course, the provision of immediate needs, such as food, clothing, and shelter. But there is also a sense of purpose that extends beyond the confines of one's own home. In what type of world will the children live, work, and raise their own families? Being a parent entails a responsibility, not simply to one's own children, but to all children. This is particularly clear when listening to the parents of those men and women killed in unjust wars and other acts of violence.

This book is written with a purpose. And through the pages that follow, I deliberately and explicitly seek to change attitudes and ideals. Although not a pacifist in the strict sense of the term, I do remain skeptical of war as a form of politics. And I am most concerned—and repulsed—by the use of war as a tool for profit.

At Ashgate I thank the support, patience, and suggestions from my commissioning editor, Valerie Rose. I am grateful for her understanding of the motivations behind this book, and her desire to produce a critical political geography of the business of war. Lastly, I am grateful for the careful read and hard work of my desk editor at Ashgate, Gemma Lowle.

Thanks are extended to the many colleagues and friends who have supported me over the years: Curt Roseman, Michael Dear, Stuart Aitken, Gary Peters, Laura Pulido, Richard Wright, Jennifer Wolch, Scott Sheridan, and Keith Collins. I am especially thankful for the encouragement and insights provided by Colin Flint and Derek Gregory. Students also have provided inspiration, understanding, and friendship. In particular, I acknowledge the advice and critical commentary (even if unasked!) offered by Steve Butcher, Sutapa Chattopadhyay, Ken Hampton, Josh Inwood, Rob Kruse, Olaf Kuhlke, Aron Massey, Steve Oluic, Gabe Popescue, Andrew Shears, Mary Swalligan, and Stacey Wicker.

I thank also my family for their continued support: Dr. Gerald Tyner, Dr. Judith Tyner, Floris Tyner, Karen Owens, and Bill Owens. And to my brother, David Tyner, I warmly dedicate this book. In some ways he is perhaps more critical of the current state of affairs in the world; I hope that this book expresses his concerns. As always, I thank my late-night companions: Bond (my puppy) and Jamaica (my cat). I like to think that they actually support my writing and re-writing, rather than simply waiting and hoping for late-night snacks.

My greatest debt remains with Belinda. The last months of completing this book have been especially hectic—and not for the usual reasons of teaching classes, meeting deadlines, and honoring prior commitments that keep me away from my family. Rather, our family has happily grown by one with the addition of Anica Lyn. Our newest daughter is four years old and is quite the handful. And over these months Belinda has provided the strength and inspiration to see our family through these remarkable and wonderful times of change and adjustment. It has been Belinda's support and encouragement that has enabled me to complete this project.

And last, I thank our daughters, Jessica and Anica. Jessica herself is now almost four years old, and continues to enlarge her world through the magic of language. Both Jessica and Anica are avid readers (okay, mostly 'listeners' at this point). But both girls enjoy the wonders of books and the imagination afforded by words. My final hope is that this book will provide both Jessica and Anica some understanding of the world they will inherit, and to inspire in them the courage to transform the world for social justice. In short, to help future generations of children live in times of peace.

Chapter 1

George Orwell's Footsteps

In March 1990, on the occasion of Terry Anderson's[1] fifth year in captivity, Senator Daniel Patrick Moynihan (D-NY) quoted French journalist and former hostage Jean-Paul Kauffman who wrote of the plight of Anderson: "The truth is that the hostages in Lebanon today have become the damned of the West. Without hope of being saved, imprisoned in silence and darkness, deprived of the sight of the world of the living, forgotten, they no longer represent anything... The most tragic thing is that this torment is administered as much from the outside by countries and people indifferent to their fate as on the inside by their captors." Senator Moynihan continued that "The plight of the hostages in Lebanon requires our constant care and attention. We must do all we can to convey to Terry Anderson and the others that we have not forgotten, that we are not indifferent, and that we will not stop fighting for their release until they are resting safely in their homes." The senator concluded that "Captivity is a day-to-day ordeal for them; it has to be for us as well."[2]

Terry Anderson, arguably the most 'well-known' American hostage, was not the first to be abducted. Neither was he the last. In Occupied Iraq a number of Americans, such as Roy Hallums, have been abducted. And Americans have not been singled out. Workers from Sri Lanka, Nepal, the Philippines and dozens of other states, have been abducted. As of April 2006, more than 425 foreigners, and untold thousands of Iraqis, have been abducted following the 2003 invasion of Iraq.[3] Some of these hostages, such as Nicholas Berg and Kim Sun-il have been violently executed. Others, including Hallums, Jill Carroll, Angelo de la Cruz and Roberto Tarongoy—the latter men both of the Philippines—have been released.

In the writing of this book I reflect deeply on the words of Kauffman and Moynihan. Hostages, regardless of nationality, are not to be forgotten. Through their personal ordeals—as well as those of their families, friends, and other associates—

---

1 Terry Anderson, former chief Middle East correspondent for the Associated Press, was abducted in Beirut on 16 March 1985. Along with several other United States citizens, Anderson remained in captivity until his release on 4 December 1991. He was the longest held American hostage.

2 "Terry Anderson Begins Sixth Year of Captivity," *Congressional Record* (Senate, 20 March 1990), page S2709. [http:www.fas.org/irp/congress/1990_cr/s900320-anderson.htm] (8/12/2005).

3 Jonathan Finer, "Journalist Jill Carroll Freed by Her Captors in Baghdad," *The Washington Post*, 31 March 2006 [http:www.washingtonpost.com/wp-dyn/content/article/2006/03/30/AR2006033000225_pf] (4 April 2006).

they become political pawns, often in geopolitical chess games that are played out far beyond their immediate situations. And it is on this score that I disagree, in part, with Kaufmann. Hostages do continue to 'represent' something. What that is, and how it is manifest, is the partial subject of this book. Hostages—and especially those perceived as foreigners—in Occupied Iraq literally embody the juncture and disruptures of neoliberal discourses and transnational, globalist practices. Their lives remind us, through tragic circumstances, that debates over, for example, neoliberalism, transnationalism, and globalization, are not merely academic debates.

The Business of War is not, however, a chronology of abductions in Iraq. Rather, it is a wide-ranging narrative of military neoliberalism and neoconservativism, of transnationalism and globalization. I write from a particular position, namely my opposition and indeed revulsion to events in Occupied Iraq. My story, therefore, is not a detached account of an invasion and occupation, but instead (and unabashedly) a political geographic polemic against the atrocities of a modern-day colonial war. In so doing I want neither to capitalize nor to minimize the lives of Roy Hallums, Kim Sun-il or the untold other hostages whose lives—and those of their families and friends—have been irrevocably altered through the war in Iraq and the broader war on terror. I agree with Slavoj Zizek that our "focus should be on what actually transpires in our societies, on what kind of society is emerging *here and now* as the result of the 'war on terror.'[4] Critical discussion and reflection, however, offer the potential that new discursive meanings may emerge that will link, in the words of Manfred Steger, "to a progressive political tradition that seeks to give institutional expression to a more democratic and egalitarian social order."[5]

War and Bodies

The war in Iraq was (and remains) a discursive war. By this I do not deny the materiality of the conflict; nor do I discount the death and destruction that surrounds Iraq. Wars are bloody and wars are violent. But wars are not inevitable. Wars are manufactured through the policies, practices, and decisions of individuals. Paul Reuber considers the means in which the violence perpetrated—by all 'sides'—in Iraq has been legitimized. He writes that geopolitical dualizations and divisions are the weapons in this war of discourse. Moreover, those who designate 'rogue states' and put them on a map precisely establish the discursive representation and legitimation for going to war against such countries.[6]

Ian Lustick likewise maintains that the conflict in Iraq was a struggle not simply economic or diplomatic or political, but discursive. He writes: "Washington describes the mission of American troops as 'liberation.' The Muslim and Arab

---

4   Slavoj Zizek, *Iraq: The Borrowed Kettle* (New York: Verso, 2004), 19.

5   Manfred B. Steger, *Globalism: Market Ideology Meets Terrorism* 2nd ed., (Lanham, MD: Rowman & Littlefield, 2005), 153.

6   Paul Reuber, "The Tale of the Just War–A Post-Structuralist Objection", *The Arab World Geographer* 6 (2003): 44-6; at 44.

world largely sees 'conquest'. Washington speaks of 'administration', 'nation-building', and 'liberation'... the Muslim and Arab world mostly hears 'exploitation' and 'occupation'. Washington describes 'victory'. The Muslim and Arab world feels humiliation and defeat. Washington will talk of 'terrorism' and 'war crimes', but the Muslim and Arab will see 'heroism' and 'resistance.'"[7]

This point was driven home in the summer of 2005 when the US State Department declared in a public demonstration of political spin-maneuvers, that the United States was no longer fighting a global war on terror but instead was waged in a "global struggle against violent extremism." According to General Richard Myers, chairperson of the Joint Chiefs of Staff, the phrase 'war on terrorism' was objectionable because "if you call it a war, then you think of people in uniform as being the solution." Steven Hadley, national security advisor, said that "It is more than just a military war on terror. It's broader than that. It's a global struggle against extremism. We need to dispute both the gloomy vision and offer a positive alternative."[8]

This candid display of rhetorical distancing is nothing less than an attempt to align current neoliberal discourse with existing material practices. As casualties mount, and the prospects of ever 'winning the peace' fade in the sands of Iraq, approval ratings for US President George W. Bush continue to fall. But rather than changing its practices—such as the occupation of Iraq—and actually concentrating on the threat of terrorism, the government 'stays the course' and, subsequently, enacts a discursive shift. Terrorism is no longer the problem, but instead violent extremism is. And the continued resistance to US and Coalition forces in Iraq may more readily be subsumed under a discourse of 'extremism' than 'terrorism'. As Eric Schmitt and Thom Shanker write in *The New York Times* in reference to the slogan-change, "New opinion polls show that the American public is increasingly pessimistic about the mission in Iraq, with many doubting its links to the counterterrorism mission. So, a new emphasis on reminding the public of the broader, long-term threat to the United States may allow the administration to put into broader perspective the daily mayhem in Iraq and the American casualties."[9] Reuber is blunt in his assessment: "The logic of war is nothing but a socio-spatial construction and convention, which lacks a basis in essentiality and never will find one."

Wars are discursive, yes, but they are also composed of bodies. Bodies that kill and bodies that die. Bodies that are abducted and bodies that are executed. Despite technological and communication advances, wars remain corporeal events that are, quite simply, de-humanizing. Discourses are constructed to provide meaning for the loss of life. Presidents and generals call on soldiers and civilians—and the lines are

---

7   Ian S. Lustick, "America's War and Osama's Script," *The Arab World Geographer* 6 (2003): 24-6; at 24.

8   Eric Schmitt and Thom Shanker, "US Officials Retool Slogan for Terror War," *The New York Times* July 26, 2005 http://www.nytimes.com/2005/7/26/politics/26strategy.html (7/27/05).

9   Schmitt and Shanker, "Officials Retool".

increasingly blurred with privatization—to sacrifice their lives in defense of a greater goal. Freedom perhaps; or maybe democracy. When President Bush announced that the invasion of Iraq had begun, he stated that the 'sacrifice' of those who were to enter conflict did so with the "respect of the American people." This war, he declared, was entered "reluctantly" but, having arrived, would be conducted with "decisive force". In the end, Americans would "defend" their "freedom" and likewise "bring freedom to others."[10] Bush identified innocents and enemies, civilians and combatants in his speech. In the months ahead new 'bodies' would enter the American public's political lexicon: hostages, insurgents, and extremists.

Derek Gregory explains that the war on terror, and its manifestation in Afghanistan, Iraq, and beyond, "is an attempt to establish a new global narrative in which the power to narrate is vested in a particular constellation of power and knowledge within the United States of America." He vividly shows "how ordinary people have been caught up in its violence: the thousands murdered in New York City and Washington on September 11, but also the thousands more killed and maimed in Afghanistan, Palestine, and Iraq under its bloody banners." Gregory's imperative is to demonstrate that "the colonial present is not produced through geopolitics and geoeconomics alone, through foreign and economic policy set in motion by presidents, prime ministers and chief executives, the state, the military apparatus and transnational corporations." On the contrary, it "is also set in motion through mundane cultural forms and cultural practices that mark other people as irredeemably 'Other' and that license the unleashing of exemplary violence against them."[11] Gregory's words evoke the haunting poetics of Edward Abbey: "The world is wide and beautiful. But almost everywhere, everywhere, the children are dying."[12] And so are their mothers and fathers, brothers and sisters.

The body, literally, is in crisis.

Political Subjects

I write from a post-colonial and post-structural perspective. My aim is to examine the political subjugation of hostages within Occupied Iraq as a means of articulating the de-humanization of neoliberalism and the business of war. Occupied Iraq constitutes a contested place where the conjunction of neoliberal discourse and transnational practices converge and diverge. I take my lead from the work of Kathryn Mitchell (among others). In her work on neoliberalism in Vancouver, Mitchell is interested in both the actual, material implications of transnational flows and networking, but

---

10 Office of the Press Secretary, "President Bush Addresses the Nation," March 19, 2003, www.whitehouse.gov/news/releases/2003/03/print/20030319-17.html (April 12, 2004).

11 Derek Gregory, *The Colonial Present: Afghanistan, Palestine, Iraq* (Malden, MA: Blackwell Publishing, 2004), 16.

12 Edward Abbey, *Abbey's Road* (New York: Penguin Books, 1972), 97.

also in the hegemonic production and resistance to an ideology of world market domination, otherwise known as neoliberalism.[13]

Hostages, certainly within Occupied Iraq, are de-humanized subjects. And it is for this reason that I follow in the footsteps of George Orwell. Throughout his brief life, Orwell's project was to understand the place of humanity within de-humanizing, totalitarian worlds. His interest was in 'who we are' and 'how we come to be'. Although appropriated by adherents from both the Right and the Left of the political spectrum, Orwell was less concerned about the ideological orientation of abuses of power as he was about the manifestation of power and the loss of humanity—of self.

Orwell understood also that the constitution of self was situated in place. Timothy Oakes, relatedly, sees 'place' as consisting of two key components: as a site of both meaningful identity and immediate agency. He elaborates that place, as a site of meaningful action for people, is action derived from linkages across space and time; this, consequently, makes place more of a dynamic web than a specific site or location.[14] The place of Occupied Iraq, as subsequent chapters demonstrate, is a dynamic web of transnational actors—military troops, privatized security personnel, contract workers, corporate officers, and political up-starts—all vying to capitalize on war.

The work of Orwell resonates with that of Michel Foucault. Associated with post-structuralism, Foucault's overall project was "to create a history of the different modes by which ... human beings are made subjects."[15] As such, much of Foucault's early writings addressed the disciplinary techniques by which bodies were discursively made. Foucault's conception of a 'politico-anatomy' of the body, especially, has provided impetus for a number of studies on body politics. In *Discipline and Punish*, for example, Foucault explains that "the body is also directly in a political field; power relations have an immediate hold upon it; they invest it, mark it, train it, torture it, force it to carry out tasks, to perform ceremonies, to emit signs."[16] As I argue in Chapter 4, the bodies of 'hostages' likewise are invested with meanings; they are literally tortured, but they are also expected to carry out political tasks and to perform public ceremonies. Abducted bodies become contested signs and spectacles for global consumption. Foucault continues that the "political investment of the body is bound up, in accordance with complex reciprocal relations with its economic use; it is largely as a force of production that the body is invested with relations of power and domination; but, on the other hand, its constitution as

---

13 Kathryn Mitchell, *Crossing the Neoliberal Line: Pacific Rim Migration and the Metropolis* (Philadelphia: Temple University Press, 2004), 12.

14 Timothy Oakes, "Place and the paradox of modernity," *Annals of the Association of American Geographers* 87: 509-31; at 510.

15 Michel Foucault, "The subject and power," in J. Faubion (ed.) *Power: Essential Works of Foucault, 1954-1984*, Volume 3, trans. R. Hurley and others (New York: The New Press, 2000), 326-348; at 326.

16 Michel Foucault, *Discipline and Punish: The Birth of the Prison*, trans. A. Sheridan (New York: Vintage Books, 1979), 25.

labour power is possible only if it is caught up in a system of subjection." In short, the body becomes "a useful force only if it is both a productive body and a subjected body."[17]

Presently, and following Foucault, I employ the term 'subject' rather than 'identity.' As Catherine Belsey explains, the term 'subject' first places the emphasis squarely on the language we learn, and from which we internalize the meanings our society expects us to live by. Second, 'subject' reflects its own ambiguities, its double meanings.[18] As Foucault writes, there are two meanings of the word 'subject': "subject to someone else by control and dependence, and tied to his [sic] own identity by a conscious or self-knowledg. Both meanings suggest a form of power that subjugates and makes subject to."[19] Third, 'subject' allows for discontinuities and contradictions, whereas 'identity' implies sameness; subjects, however, can differ—even from themselves.

This notion of the 'subject' thus differs from the more traditional, Cartesian-based 'individual' of the Enlightenment. The latter concept, 'individual', dates from the Renaissance and presupposes that humans are free, intelligent entities, and that decision-making processes are not coerced by historical, political, or cultural circumstances. As Edkins explains, the "Enlightenment subject was a unified individual with a center, an inner core that was there at birth and developed as the individual grew, while remaining essentially the same. This core of the self was the source of the subject's identity."[20] The poststructural subject, conversely, has no fixed, essential, or permanent identity. "Subjectivity is formed and transformed in a continuous process that takes place in relation to the ways we are represented or addressed and alongside the production or reproduction of the social."[21]

Foucault forwards the idea that "the individual is not a pre-given entity which is seized on by the exercise of power. The individual, with his [sic] identity and characteristics, is the product of a relation of power exercised over bodies, multiplicities, movements, desires, forces."[22] Foucault thus does not deny the materiality of the body; but neither does the body's materiality exist outside a disciplinary framework in terms of both knowledge and practices.[23] Language therefore is not a tool to express ideas about reality; rather, the speaking subject is always already embedded in a preexisting language structure. Naming produces

---

17 Foucault, *Discipline and Punish*, 25-26.

18 Catherine Belsey, *Poststructuralism: A Very Short Introduction* (Oxford: Oxford University Press, 2002), 52.

19 Foucault, "The subject and power," 331.

20 Jenny Edkins, *Poststructuralism and International Relations: Bringing the Political Back In* (Boulder, CO: Lynn Rienner Publishers, 1999), 21.

21 Edkins, *Poststructuralism and International Relations*, 22.

22 Michel Foucault, "Questions on Geography," in C. Gordon, *Power/Knowledge: Selected Interviews and Other Writings, 1972-1977* (New York: Pantheon Books, 1980); 63-77; at 73-74.

23 M.A. McLaren, *Feminism, Foucault, and Embodied Subjectivity* (Albany: State University of New York Press, 2002), 15.

things rather than attaching labels to 'objects' that are already there.[24] Political language, in short, attempts to produce its own subjects.

Knowledge/power/discourse

Knowledge about any given object, and especially the production of that knowledge, is crucial to the understanding of political practices, including the subjugation of bodies. In the many writings of Foucault, knowledge assumes a central place. Indeed, a dominant motif of Foucault's work was to provide a critique of the way modern socities control and discipline their populations by sanctioning the knowledge claims and practices of the human sciences.[25] Spokespersons for the US Department of State, for example, routinely couch their briefings in terms of the *known* and the *unknown*. Indeed, it was US Secretary of Defense Donald Rumsfeld who remarked: "As we know, there are known knowns; there are things we know we know. We also know there are known unknowns; that is to say we know there are some things we do not know. But there are also unknown unknowns—the ones we don't know we don't know."[26]

For Foucault, knowledge is inseparable from power. Foucault explains that power and knowledge directly imply one another; there is no power relation without the correlative constitution of a field of knowledge, nor any knowledge that does not presuppose and constitute at the same time power relations.[27] Power, in a Foucauldian sense, is thus very different from most accounts. Power is not something that is possessed, or the sole domain of a select few individuals or institutions. Rather, power is exercised. Foucault elaborates that "Power must be analysed as something which circulates, or rather as something which only functions in the form of a chain. It is never localised here or there, never in anybody's hands, never appropriated as a commodity or piece of wealth."[28] More on this will be said in Chapter 4.

A focus on power/knowledge relations leads to an engagement with discourses. Discourse, for our present discussion, refers not the standard dictionary definition of a conversation or written expression, but rather to *disciplines of knowledge*. In *The Archaeology of Knowledge*, Foucault outlines the coordinates of discourse and discursive formations. Discourses are derived from statements which, in turn, have four attributes. First, statements must have a material existence; they must be produced, circulated, and consumed. The material manifestation of statements may include oral or written enunciations; however, statements may also appear as

---

24 Edkins, *Poststructuralism and International Relations*, 22.

25 Sarup, *Post-structuralism*, 72.

26 "Rumsfeld Remark Win Rumsfeld an Award," *BBC News*, 2 December 2003 [http://news.bbc.co.uk/go/pr/fr/-/2/hi/americas/3254852.stm] (4 April 2006).

27 Foucault, *Discipline and Punish*, 27.

28 Michel Foucault, "Two lectures," in C. Gordon (ed.) *Power/Knowledge: Selected Interviews and Other Writings, 1972-1977*, trans. C. Gordon, L. Marshall, J. Mepham, and K. Soper (New York: Pantheon Books, 1980), 78-108; at 98.

graphs, tables, maps, photographs, and so forth. Second, statements must have a substance: they are manifest in particular places and at specific times. This suggests that statements are both contextual and contingent. Third, statements do not have as their correlate an individual or a particular object; in other words, there is no essential referent to a statement. Instead, statements—and by extension, discourses—form the objects of which they speak. Last, statements are always bordered by other statements. There is no statement, according to Foucault, that is free, neutral, or independent; instead, a statement always belongs to a series or a whole, always plays a role among other statements, deriving support from them.[29] Combined, a collectivity of statements constitutes a discursive formation; discursive formations, in turn, produce bodies of knowledge.

Having established the parameters of statements and the notion of discursive formations, Foucault forwards a number of principles that guide his understanding of discourses. First, Foucault suggests that discourses must be treated as discontinuous. Second, Foucault details a principle of rarity, arguing that 'everything is never said'. All knowledges are partial and, consequently, discourses are selective. Third, there exists a principle of specificity, which states that a particular discourse cannot be resolved by a prior system of significations; in other words, we should not imagine that the world presents us with a legible face, leaving us merely to decipher it. Last, discourse finds a way of limiting its domain, of defining what it is talking about, of giving it the status of an object—and therefore of making it manifest, nameable, and describable. A Foucauldian approach is to view discourses not as groups of signs but as practices that systematically form the objects of which they relate. Consequently, the knowledge that enables us to answer questions pertaining to a particular field is informed by the discursive practice constituting and demarcating the field.[30]

We have come full circle. Bodies of knowledge are socially produced. More crucial, however, is the assertion that these knowledges produce that of which they speak. There is no essential truth waiting to be discovered. All meaning, all interpretation, is contingent and socially-defined. Knowledge, simply put, is political discourse. In the remainder of this chapter I highlight six concepts—neoliberalism, neoconservatism, militarism, globalization, transnationalism, and security—that guide my argument throughout the course of this book. These six concepts provide the reference points for the invasion and occupation of Iraq as well as the subsequent abductions and killing of hostages.

---

29 Michel Foucault, *The Archaeology of Knowledge & the Discourse on Language*, trans. A. Sheridan Smith (London: Pantheon Books, 1972), 99.

30 N. Gordon, "On visibility and power: an Arendtian corrective of Foucault," *Human Studies* 25(2002): 125-45; at 127.

Neoliberalism

Capitalism is an economic system in which all economic actors—producers and consumers—depend on the market for their basic needs.[31] Waged labor is a defining characteristic, as is private ownership of the means of production (e.g., capital, land, and labor). Capitalism is driven by certain systemic imperatives, namely the imperatives of competition, profit-maximization, and profit-accumulation.[32] Indeed, as David Harvey explains, accumulation is the engine which powers growth under the capitalist mode of production.[33]

Capitalism, however, is also inherently and inevitably crisis-prone. This occurs because economic growth under capitalism is a process of internal contradictions. The drive for profit makes capitalism both relentlessly expansive and prone to contradiction, both stemming from a dynamic process of over-accumulation.[34] Termed variously as a *crisis of over-accumulation, overproduction*, or *overcapacity*, this crisis is marked by chronic unemployment and underemployment, capital surpluses and lack of investment opportunities, falling rates of profit, and a lack of effective demand in the market.[35] Confronted with a crisis of over-accumulation, states may intervene in any number of ways. Of particular importance for our present discussion is a dynamic David Harvey terms a 'spatial fix'. Through this process, excess capital and labor may be absorbed through geographic expansion. Capitalism is 'spread' to new territories through trade, foreign investment, and the outsourcing of employment.

Most recently, this spatial fix, in the American context, has found expression in the promotion of neoliberalism.[36] Rooted in classical liberal ideals of British

---

31 Ellen Meiksins Wood, *Empire of Capital* (London: Verso, 2003), 9; see also Richard Peet, *Global Capitalism: Theories of Societal Development* (New York: Routledge, 1991).

32 Wood, *Empire of Capital*, 10.

33 David Harvey, *Spaces of Capital* (New York: Routledge, 2001), 237.

34 Although crises of capitalism may appear in many ways, it is useful to consider one particular situation. In order to accumulate capital (profit maximization), a factory owner will pay the least amount in wages to his or her employees. Also, in light of competition from other factories, workers are encouraged to produce more and more goods. However, with reduced wages, workers are unable to consume at a rate commensurate with the production of goods. This leads to a situation wherein there is an overproduction of capital and/or an underconsumption of products. For further discussions on crises of overaccummulation, see Harvey, *Spaces of Capital*, 2001; David Harvey, *The Limits to Capital* (Oxford: Basil Blackwell, 1982); David Harvey, *The Condition of Postmodernity* (Oxford: Basil Blackwell, 1989).

35 Walden Bello, *Dilemmas of Domination: The Unmaking of the American Empire* (New York: Henry Holt and Company, 2005), 4; Harvey, *Spaces of Capital*, 240.

36 The central tenets of neoliberalism include the primacy of economic growth, the importance of free trade to stimulate growth, the unrestricted free market, individual choice, the reduction of government regulation, and the advocacy of an evolutionary model of social development anchored in the Western experience and applicable to the entire world; see Steger, *Globalism*, 8-9.

philosophers, neoliberalism premises that states should play a minimal role in the day-to-day workings of markets. Liberalism, as a critique of feudalism, emphasized the priority of the individual and thereby privileged concepts such as reason, equality, and competition. This contrasted with earlier feudalistic societies that were defined by community, authority, and hierarchy. Liberty was especially promoted, although in this sense the term meant the relative absence of external impediments to rational behavior.

Neoliberalism traces its lineage to classical liberal thinkers, including Adam Smith (1723-1790), David Ricardo (1772-1823), and Herbert Spencer (1820-1903). Smith, who is generally credited as developing the ideas of capitalism, asserted that a nation's wealth was not determined by the amount of gold in its treasury—as typified by mercantilism—but rather wealth was determined by productivity. Accordingly, Smith forwarded the principle of laissez-faire capitalism, which suggested that governments should remain separate from economic activities. He presumed that the resources of a nation would be best managed when individuals were allowed to pursue economic activities according to market principles. In particular, Smith reasoned that the 'invisible hand' of supply and demand would assure the most beneficial functioning of production and consumption activities.[37]

Capitalism, as it developed, was oriented toward the individual. As opposed to feudalistic or communal societies, capitalism assumes that society's best interest is maximized when each individual is free to do that which he or she thinks is best for himself or herself.[38] Only through the self-determination of people freely operating within markets would society as a whole benefit. Consequently, Smith stressed the 'unseen' work of the markets, which not only capitalized on individual freedoms and organized for the provision of mass needs, but did so in such a way that the overall interests of the commonwealth were supposedly safeguarded and advanced.[39] The market came to be seen as a self-regulating mechanism tending toward equilibrium of supply and demand, thus securing the most efficient allocation of resources.[40] Smith concluded that "A society of individual property owners, free and independent before the law and with equal recourse to it, who met in the competitive marketplace, and who enjoyed the right to a democratic vote and voted unashamedly in their own self interest—such a society, the Enlightenment promised, would produce the best outcome for everyone."[41]

These ideas were expanded through the work of Ricardo. He suggested that it was appropriate for capitalists to force labor to surrender a large part fo the value created; this exploitation was necessary to ensure the continued existence of private capital

---

37   Leon P. Baradat, *Political Ideologies: Their Origins and Impact* 7th ed., (Upper Saddle River, NJ: Prentice Hall, 2000), 86.

38   Baradat, *Political Ideologies*, 86.

39   Neil Smith, *The Endgame of Globalization* (New York: Routledge, 2005), 31.

40   Steger, *Globalism*, 9.

41   Smith, *Endgame*, 32.

that would be needed for future investment.[42] This formed the backbone of Ricardo's theory of the Iron Law of Wages wherein capitalists should only pay workers enough to keep them returning to their job. Capitalism likewise required an environment whereby other non-wage modes of existence were removed. In short, capitalist to be fully profitable required the participation of everyone—willing or not.

Ricardo further developed the idea of comparative advantage. This was used as argument against government interference with free trade. First specified by Robert Torrens (1780-1864) in 1815, the law of comparative advantage states that a country will benefit by exporting a good that it produces at a lower relative cost than other countries; conversely, a country will benefit by importing a good that it could produce at a higher relative cost. Consequently, all countries would eventually specialize only in the production of those commodities for which it had a comparative advantage.

Competition among numerous small buyers and sellers was viewed as the engine of economic growth. This idea was elaborated by the social Darwinian writings of Herbert Spencer. Spencer's liberalism was concerned with the welfare of society, but from an economically competitive orientation. Modifying the evolutionary ideas of Charles Darwin, Spencer argued that welfare programs were tampering with the 'invisible hand' of evolution. In so doing, Spencer argued that free-market economies constituted the most civilized form of human competition in which the 'fittest' would 'naturally' rise to the top. Consequently, he denounced socialism, trade unions, and social regulation in so far as these, according to Spencer, would inhibit rational progress and individual freedom.[43] Spencer's theory was immensely popular in the United States, resonating strongly with the 'rugged individualism' that was promoted in America.

Given this legacy, the central features of neoliberalism include the primacy of economic growth, the importance of free trade to stimulate growth, the unrestricted free market, individual choice, privatization of business, the reduction of government regulation, and the advocacy of an evolutionary model of social development anchored in the Western experience and applicable to the entire world.[44] Harvey summarizes that neoliberalism is, first, a theory of political economic practices that proposes that human well-being can best be advanced by liberating individual entrepreneurial freedoms and skills within an institutional framework characterized by strong private property rights, free markets, and free trade. Furthermore, if markets do not exist, then they must be created, by state action if necessary.[45]

Proponents of neo-liberalism believe that only through the constant expansion of capitalism will the problems of the world be eliminated. Critics, however, argue that it is capitalist accumulation itself that is the root of mass poverty, grotesque inequalities in living conditions, and so on. Ignacio Ramonet explains that "In the

---

42 Baradat, *Political Ideologies*, 87.
43 Steger, *Globalism*, 10.
44 Steger, *Globalism*, 8-9.
45 David Harvey, *A Brief History of Neoliberalism* (Oxford: Oxford University Press, 2005), 2.

era of neoliberalism, geopolitical supremacy and the actions of the hyperpower have proven incapable of guaranteeing a satisfactory level of human development for all citizens. Among the inhabitants of a country as rich as the United States, for example, there are 32 million people whose life expectancy is less than 60 years; 40 million without medical coverage; 45 million living below the poverty line; and the figure for functional illiteracy is 52 million."[46]

Henry Giroux maintains that "We live at a time when the conflation of private interests, empire building, and evangelical fundamentalism brings into question the very nature, if not the existence, of the democratic process." He argues that the social contract is under attack, with its emphasis on enlarging the public good and expanding social provisions—such as access to adequate health care, housing, employment, public transportation, and education—which have provided a limited though important safety net. In its place has emerged a notion of national security based on fear, surveillance, and control. Indeed, under a neoliberal domestic restructuring, militant foreign policy, and evangelical conservatism, the United States has witnessed the increasing obliteration of those discourses, social forms, public institutions, and noncommercial values that are central to the language of public commitment, democratically charged politics, and the common good.[47] Giroux concludes that "With its debased belief that profit-making is the essence of democracy, and its definition of citizenship as an energized plunge into consumerism, neoliberalism eliminates government regulation of market forces, celebrates a ruthless competitive individualism, and places the commanding political, cultural, and economic institutions of society in the hands of powerful corporate interests, the privileged, and unrepentant religious bigots."[48] And lastly, Aihwa Ong describes American neoliberalism as "an extreme realization of the priority of market principles, which are now invading all areas of social life and exposing citizens to levels of risk from which they have heretofore been partially protected. Its logic entails a sustained assault on democratic institutions, such as the welfare state and labor unions, that traditionally serve as countervailing powers vis-à-vis market forces." She continues that American neoliberalism, by excessively privileging individual rights, undermines democratic principles of social equality."[49]

Neoconservatism

Frequently, neoliberalism has been coupled with another discourse, neo-conservatism. In certain respects, these two ideas are variations on the same liberal themes: the

---

46 Ignacio Ramonet, *Wars of the 21st Century: New Threats, New Fears*, trans. Julie Flanagan (New York: Ocean Press, 2004), 6.

47 Henry A. Giroux, *The Terror of Neoliberalism* (Boulder, CO: Paradigm Publishers, 2004), xv-xvi.

48 Giroux, *Terror of Neoliberalism*, xvii.

49 Aihwa Ong, *Flexible Citizenship: The Cultural Logics of Transnationality* (Durham, NC: Duke University Press, 1999), 211-2.

importance of free markets and free trade. However, neo-conservatives are more apt to combine a hands-off approach to economic regulation with intrusive government action for the regulation of ordinary citizens in the name of public security and traditional values.[50] In other words, governmental interference in social spheres is acceptable, presuming that this interference, first, promotes a particular morality and value system and, second, that it does not alter the free market system. Neo-conservatives, accordingly, do not favor social welfare programs; instead, Spencerian ideas of private responsibility and individualism are promoted.

According to Andrew Bacevich, neoconservatism originated out of the fall-out of America's defeat in Vietnam. Various intellectuals believed that the lessons of Vietnam did not revolve around the over-extension of American power, nor around a moral irresponsibility. Rather, the consequences of defeat in Vietnam to these individuals demonstrate the absence of American power and will. Such weaknesses endangered the security and prosperity of the United States and its allies.[51]

The neoconservative movement in the United States, as it emerged through the 1970s, viewed state power not as a necessary evil, but as a positive good to be cultivated and deployed. It was, in essence, a transformative endeavor to remake American society and, by extension, the global environment. To this end, Bacevich identifies six propositions that summarize the essence of the neoconservative movement. First, there is a fundamental understanding of history that is built upon to 'truths'. The first is that 'evil' is real; the second is that for evil to prevail, it requires only one thing: for those confronted by evil to flinch from duty. Bacevich argues that this understanding of history owes its existence to a reading of the Great Depression of the 1930s and the rise of Adolf Hitler and Nazi Germany.[52]

A second proposition stems from this concern with the Second World War, namely that diplomacy, accommodation, and such did not prevent Nazi Germany from the path of aggression. What was required was an effective use of superior force. The lesson for neoconservatives was that "in international politics there was no substitute for power, especially military power."[53]

A third proposition relates to the perception of America's 'mission'. Reworking the idea of Manifest Destiny, neoconservatives believe that alternatives to or substitutes for American global leadership simply do not exist. As Bacevich writes, according to neoconservatives, "History had singled out the United States to play a unique role as the chief instrument for securing the advance of freedom, which found its highest expression in democratic capitalism. American ideals defined America's purpose, to be achieved through the exercise of superior American power."[54] Here we see clearly the confluence of neoliberal economic ideas and the moral justifications

---

50   Steger, *Globalism*, 16.

51   Andrew J. Bacevich, *The New American Militarism: How Americans are Seduced by War* (Oxford: Oxford University Press, 2005), 70.

52   Bacevich, *New American Militarism*, 73.

53   Bacevich, *New American Militarism*, 73-74.

54   Bacevich, *New American Militarism*, 75.

for American empire-building. The neoconservative moment thus re-captures many of the claims to legitimacy that were used in support of earlier American military interventions, including the colonization of the Philippines, Guam and Puerto Rico.

Vietnam provided another lesson for neoconservatives, namely that foreign policy must not be subsumed by domestic policies. The fourth proposition, therefore, concerns the relationship between cultural politics and America's purpose abroad. Simply put, neoconservatives promote 'traditional' values, include marriage and the nuclear family, the advocacy of law and order, and respect for organized religion. Only by drawing on these beleaguered institutions, it was maintained, could the United States promote a powerful and transformative foreign policy. In short, "only by ensuring order and stability at home and restoring confidence in basic institutions ... could the United States fend off the Communist threat and fulfill the historical mission for which it had been created."[55]

A fifth proposition, also drawing on Vietnam, suggests that absent decisive action to resolve a crisis, unspeakable consequences await. And for neoconservatives, crisis is a permanent condition. Reflecting a heightened form of *realpolitik*, neoconservatives view the world as in a state of total and unending risk. Within such an environment, a powerful military is required, as well as a political leader who is willing to use such force. Such is the context, as subsequent chapters illustrate, for the foundation for America's 'shock and awe' campaign in Iraq and the birth of the Bush Doctrine. Moreover, the 'War on Terror' provides a convenient umbrella to legitimate the discourse of permanent crisis.[56]

Lastly, and in effect an off-shoot of the previous propositions, neoconservativism maintains that strong leadership is required to guide America to its global destiny. As Bacevich concludes, neoconservatives seek leaders who demonstrate unflinching determination, moral clarity, and inspiration: in short, neoconservatives advocate for 'heroic' leadership.[57]

As with all movements, the basic characteristics have been modified. For neoconservatism, changes resulted largely from the collapse of the Soviet Union and the end of the Cold War. For decades the specter of Communism was used as a justification for American military supremacy and global leadership. With the dissolution of the Soviet Union neoconservatives were left adrift. However, even as some writers and political pundits declared the neoconservative project dead, a second generation of neocons rose to prominence. As Bacevich writes, these neoconservatives proposed a decidedly more aggressive and transformative agenda for the United States. As the world's lone superpower, it was argued, America must rightly fulfill its destiny to remake the world in its own image. It was at this time that neoconservatives began to warmly embrace the idea of an American *empire*.[58]

---

55 Bacevich, *New American Militarism*, 76-77.
56 Bacevich, *New American Militarism*, 77.
57 Bacevich, *New American Militarism*, 77-78.
58 Bacevich, *New American Militarism*, 79-80.

Second-generative neoconservatism, according Bacevich, includes five convictions. The first is a belief that American global dominion is benign and that other nations necessarily see it as such. In essence, other states *want* America to lead the world. Second, failure on the part of the United States to assume such a role would lead to a global disorder. Third, the means to assume such leadership is through the use of force—specifically, military force. For neoconservatives, "Employing that military might with sufficient wisdom and determination could bring within reach peace, prosperity, democracy, respect for human rights, and American global primacy extending to the end of time." Fourth, such a belief suggests that American leadership must be committed to sustaining and even enhancing American military supremacy. Lastly, second-generation neoconservatives abandoned their political realism in favor of American hegemony. Realism, according to these neoconservatives, posed a problem in that it entailed a defense of national interests rather than a transformation of the global order. Accordingly, neoconservatives argue for a more pro-active and aggressive foreign policy, one that emphasizes "the use of armed force to promulgate American values and perpetuate American primacy."[59]

Neoconservatism is associated with the rise of militarism in the United States. As detailed by Harvey, "neoconservatism is ... entirely consistent with the neoliberal agenda of elite governance, mistrust of democracy, and the maintenance of market freedoms." However, according to Harvey, it departs from neoliberalism in two fundamental ways. First, "is its concern for order as an answer to the chaos of individual interests, and second, in its concern for an overweening morality as the necessary glue to keep the body politic secure in the face of external and internal dangers." Consequently, neoconservatives undemocratically stress morality, conformity, and authoritarianism in their pursuit of economic, political, and social agendas. This dovetails readily with militarism in that, as Harvey writes, "some degree of coercion appears necessary to restore order." Indeed, neoconservatives "are far more likely to highlight threats, real or imagined, both at home and abroad, to the integrity and stability of the nation."[60]

Militarism

Militarism is a crucial component in the business of war. Indeed, it is through the promotion of a militarized society that makes the business of war seem natural and normal. In recent years a number of commentators have identified the emergence of a 'new' or 'intensified' militarism within American society. Bacevich states his case bluntly: "Today as never before in their history Americans are enthralled with military power. The global military supremacy that the United States presently enjoys—and is bent on perpetuating—has become central to our national identity."[61]

---

59 Bacevich, *New American Militarism*, 83-87.
60 Harvey, *Brief History of Neoliberalism*, 82.
61 Bacevich, *New American Militarism*, 1.

The fact is, however, militarism has long been associated with the United States. It is for this reason that Michael Mann identifies a trend he terms the "new" militarism. For Mann, the "new militarism has the customary strengths and weakness of militarism—power but not authority, ruthless arrogance leading to overconfidence, eventually leading to hubris and disaster." However, what sets apart the more recent neoconservative-influenced militarism is the incoherence of American actions. He maintains that "Whereas in the recent past American power was hegemonic—routinely accepted and often considered legitimate abroad—now it is imposed at the barrel of a gun. This undermines hegemony and the claim to be a benevolent Empire. Incoherence among its military, economic, political and ideological powers forces it to retreat to its strongest resource, offensive military devastation."[62]

Before discussing militarism within contemporary American society, however, it is useful to sketch out the parameters of militarism. Rachel Woodward, for example, defines militarism as "the shaping of civilian space and social relations by military objectives, rationales and structures, either as a part of the deliberate extension of military influence into civilian spheres of life and the prioritising of military institutions, or as a byproduct of those processes."[63] Woodward's definition is useful in that it highlights the intersection of military and civilian spheres of influence. However, it tends to be rather one-side, with military objectives positioned as the determining factor. Conversely, we should consider also the shaping of military spaces by civilian institutions. More specifically, we should not lose sight that economic ideologies—including but not limited to neoliberalism and neoconservatism—have greatly impacted the role and function of the military.

Woodward's definition compliments that provided by Chalmers Johnson. Johnson describes militarism as "the phenomenon by which a nation's armed services has come to put their institutional preservation ahead of achieving national security or even a commitment to the integrity of the governmental structure of which they are a part." He continues that "when a military is transformed into an institution of militarism, it naturally begins to displace all other institutions within a government devoted to conducting relations with other nations." For Johnson, one sign "of the advent of militarism is the assumption by a nation's armed forces of numerous tasks that should be reserved for civilians."[64] This is expressed most clearly in the emergence of private military corporations (discussed below) and again points to the confluence of economic ideologies with the military.

Another definition is offered by Bacevich. In his writings, he emphasizes the impact of militarism on American society and the tendency to see military power as the ultimate arbiter. Bacevich argues, for example, that Americans have fallen prey

---

62 Michael Mann, *Incoherent Empire* (London: Verso, 2003), 252.

63 Rachel Woodward, "From Military Geography to Militarism's Geographies: Disciplinary Engagements with the Geographies of Militarism and Military Activities," *Progress in Human Geography* 29 (2005): 718-740; at 721.

64 Chalmers Johnson, *The Sorrows of Empire: Militarism, Secrecy, and the End of the Republic* (New York: Henry Holt and Company, 2004), 24.

to militarism, manifesting itself in a romaticized view of soldiers, and a tendency to see military power as the truest measure of national greatness; Americans have come to define the nation's strength and well-being in terms of military preparedness, military action, and the foster of military ideals.[65] This theme is picked up by other writers. Indeed, Clyde Prestowitz suggests that the United States "relies very heavily on one card in the international poker game, the miliary card." According to Prestowitz, recent events indicate quite clearly that America is becoming more and more dependent on the military for its conduct of foreign policy as well as for the US economy. He notes, first, that in recent years America has rejected or weakened several landmark treaties, including the ban on use of landmines, the ban on trade in small arms, the comprehensive test ban treaty, the ABM treaty, the chemical warfare treaty, the biological warfare treaty, the nonproliferation treaty, and the International Criminal Court. All of this suggests, to Prestowitz, that the US government has no use for the United Nations or other multilateral institutions and, instead, places complete faith in American power.[66]

Militarism is patently apparent in America's foreign policy. Chalmers Johnson, for example, argues that America "prefers to deal with other nations through the use or threat of force rather than negotiations, commerce, or cultural interaction and through military-to-military, not civilian-to-civilian, relations."[67] But militarism is also viewed as a crucial component of America's economy. The arms industry, as a case in point, is a major exporter, provider of jobs, and form of state leverage. Prestowitz explains that "The United States uses arms arrangements to cement relationships with key countries, to standardize equipment and procedures globally, and also to gain a degree of control over foreign government policies."

As indicated above, it is not possible to under emphasize the interconnections between militarism and neoliberalism. Iain Boal and his colleagues note that *military neoliberalism* is the key formula to a proper determination of the present capitalist moment. Military neoliberalism is, in actuality, a form of primitive accumulation. The dispossession of laborers; all forms of dispossession; and this is fundamental to capitalism as a system.

Why has capitalism adopted a specifically military (imperialist) form? Boal et. al. suggest that, on the one hand, primitive accumulation has always been an exercise in violence. On the other hand, in recent years—and especially following the decades of wars of liberation—the dominant capitalist core finds it harder to benefit from consensus market expansion. Indeed, according to Boal and his colleagues, "In the world at large—the world neo-liberalism was fighting to create—struggles

---

65 Bacevich, *New American Militarism*, 2.

66 Clyde Prestowitz, *Rouge Nation: American Unilateralism and the Failure of Good Intentions* (New York: Basic Books, 2003), 144, 161.

67 Johnson, *Sorrows of Empire*, 5.

accumulated as the price of change became clearer, and capitalism's enemies began to score real successes."[68]

In such a contested global environment, the military option—for those countries like the United States so equipped—begins to make more sense. To be sure this does not constitute a radical rupture in US foreign policy. As Boal and his colleagues assert, military interventions have served as the primary strategic element in a history of relentless imperial and capital expansion throughout the existence of the United States. Each military campaign—whether in the Philippines in 1898 or Vietnam in the 1950s and 1960s—has been designed to serve an overall strategic project of pressing American power on distant lands. In short, such military intervention has been crucial for the expansion of Western capital entrenchment in 'emerging markets.'[69]

Militarism is thus directly associated with neoliberal globalization, the rise of neoconservatism, and military-related transnational practices. And, as such, one 'firm' that has benefitted tremendously from the use of temporary workers is the United States Armed Forces. As discussed by Peter Singer, the growth of a privatized military industry is part-and-parcel of the changed global security and business environments of the twenty-first century. Associated with the emergence of neoliberal philosophies, a large number of functions previously performed by government militaries have been outsourced to private contractors. To this end, various services traditionally falling within the domain of national militaries, such as combat operations, strategic planning, military training, intelligence, military logistics, and information warfare, are being conducted by private firms.[70]

Privatized Military Firms (PMFs) or Corporations (PMCs) are not simply 'mercenary' organizations. Rather, these corporations are business organizations that trade in professional services intricately linked to warfare. They are in fact neoliberal innovations—exemplars of a militarized society. To this end, Singer raises a crucial point regarding the privatization of the military: although soliders often serve to prevent wars, private military firms require wars. This necessarily involves their casting aside a moral attitude toward war.[71]

PMCs have emerged, especially in the post-Cold War era, part and parcel with a broader transfer of public responsibilities to the private sector. Health care, police, prisons, garbage collection, postal services: these are examples of services that have been shifted back and forth between being viewed as essential public responsibilities of government to something left to the private sector. This relocation of service

---

68   Iain Boal, T.J. Clark, Joseph Matthews, and Michael Watts, *Afflicted Powers: Capital and Spectacle in a New Age of War* (New York: Verso, 2005), 74.

69   Boal et al., *Afflicted Powers*, 79-81.

70   Peter W. Singer, *Corporate Warriors: The Rise of the Privatized Military Industry* (Ithaca, NY: Cornell University Press, 2003), 73.

71   Singer, *Corporate Warriors*, 41.

provision, known as both 'outsourcing' and 'privatization', has become a reality in the business of war.[72]

In particular, the 'privatization revolution', a hall-mark of neoliberalism, provided the logic, legitimacy, and models for the entrance of markets into formerly state domains.[73] Operationally, PMFs are structured as firms and operate as businesses first and foremost. As business entities, they are often linked through complex financial ties to other firms, both within and outside their industry.[74] Indeed, PMFs are ordered along pre-existing corporate lines, usually with a clear executive hiearchy that includes boards of directors and share-holdings.[75]

Singer organizes privatized military industry into three broad sectors: military provider firms, military consultant firms, and military support firms. *Military provider firms* are defined by their focus on the tactical environment; these firms provide services at the forefront of the battle-space, by engaging in actual fighting. *Military consulting firms* provide advisory and training services integral to the operation and restructuring of a client's armed forces; services offered include strategic, operational, and/or organizational analysis. Unlike military provider firms, these firms do not operate on the battlefield. Lastly, *military support firms* provide nonlethal aid and assistance functions, such as logistics, intelligence, technical support, supply, and transportation. Military support firms are the largest in scope and revenue, and also the most varied. Crucially, these firms assume the more mundane functions of warfare, including feeding, clothing, transporting, and basically servicing troops.[76] Singer concludes that the privatized military industry is distinctly representative of the changed global security and business environment at the start of the twenty-first century.[77] As such, it is important to consider the emergence of private military firms within the larger debates surrounding globalization.

Globalization

Globalization is a much debated term. For some writers, globalization signifies a unique moment in the expansion of capitalism; critics of globalization, conversely, contend that the term represents nothing more than the latest 'fad' in academia. However, even a cursory consideration of widespread inequalities among peoples and countries, the increasing scarcity of drinking water, the rapid pace of biodiversity loss and deforestation suggests that something is amiss. Too many people are dying for 'globalization' to be merely a fad.

But what exactly is globalization, and how does it differ from another term often used interchangeably, namely 'internationalization'? Peter Dicken writes

---

72 Singer, *Corporate Warriors*, 7-9.
73 Singer, *Corporate Warriors*, 49.
74 Singer, *Corporate Warriors*, 40.
75 Singer, *Corporate Warriors*, 45.
76 Singer, *Corporate Warriors*, 92-8.
77 Singer, *Corporate Warriors*, 49.

that internationalization refers simply to the increasing geographical spread of economic activities across national boundaries; as such it is not a new phenomenon. 'Globalization' of economic activity is qualitatively different. It is a more advanced and complex form of internationalization which implies a degree of functional integration between internationally dispersed economic activities.[78] As such, Michael Carnoy and Manuel Castells define a global economy as the economy whose core, strategic activities have the technological, organizational, and institutional capacity to work as a unit in real time, or in chosen time, on a planetary scale. Globalization thus differs from processes of internationalization for one simple reason, as Carnoy and Castells explain:

> only at this point in history was a technological infrastructure available to make it possible. This infrastructure includes networked computer systems, advanced telecommunications, information-based technology, fast transportation systems for people, goods, and services, with a planetary reach, and the information processing capacity to manage the complexity of the whole system.[79]

It is also important to distinguish between globalization—a set of real historical social processes of increasing interdependence—and globalism—a political discourse endowing globalization with market norms, values, and meanings.[80] Globalization therefore is a material process. It is also "a linguistic and ideological practice—it is a persuasive story, embedded in a neoliberal political project."[81] The rhetoric of globalism is the rhetoric of world market domination and the ostracizing, if not extermination, of all forms of government based on any alternatives to this goal.[82] According to Mitchell, "It is the combination of the discursive framing of neoliberal globalization as necessary, inevitable, and beneficial, alongside the growth of global interdependencies in areas ranging from production, trade, finance and migration to cosmopolitan consciousness, ethnoscapes, the media, and culture, that makes contemporary globalization unique to this historical moment."[83]

Neoliberal globalists, according to Steger, believe in the creation of a single, global market in goods, services, and capital. They suggest that all peoples and states are equally subject to the logic of globalization, which is, in the long run beneficial and inevitable, and that societies have no choice but to adapt to this world-shaping force. Steger continues that this interpretation as being driven by the mysterious yet irresistible forces of the market is frequently expressed in quasi-religious language

---

78 Peter Dicken, *Global Shift: The Internationalization of Economic Activity*, 2nd ed. (New York: The Guilford Press, 1992), 1.

79 Michael Carnoy and Manuel Castells, "Globalization, the knowledge society, and the network state: Poulantzas at the millennium," *Global Networks*, 1 (2001) :1-18; at 3.

80 Steger, *Globalism*, 18.

81 Steger, *Globalism*, 19.

82 Mitchell, *Crossing the Neoliberal Line*, 12.

83 Mitchell, *Crossing the Neoliberal Line*, 12.

that bestows almost divine wisdom upon the market.[84] As Steger concludes, globalism constitutes a coherent discursive regime that shapes the social understandings of authority, but it is not a monolithic ideology.[85]

Transnationalism

The concept of transnationalism has generated considerable discussion with respect to population movements. In particular, the idea of simultaneous and multiple relationships to two or more countries has raised questions regarding the nature and meaning of citizenship and government regulation. Transnationalism, for Linda Basch and her colleagues, is defined as the processes by which immigrants forge and sustain multi-stranded social relations that link together their societies of origin and settlement. They maintain that transnationalism is distinctive in that many immigrants today build social fields that cross geographic, cultural, and political borders. Accordingly, *transmigrants* develop subjectivities embedded in networks of relationships that connect them simultaneously to two or more states.[86] Although this conception of transnational is important for my later discussion of hostages, in *The Business of War* I am primarily concerned with the processes underlying transnational labor migration as it is associated with the emergence of neoliberalism, neoconservatism, militarism, and globalization. Transnationalism is therefore viewed as a material practice, but one that is greatly informed by ideology. Transnationalism, in this context, thus refers to any practice that entails the movement of capital, commodities, or people between nation-states.

Ellen Meiksins Wood explains that capitalism, particularly as it derived from the ideas of Ricardo, depends on expropriation: States must ensure that those without capital are available as labor. This is a delicate balance. On the one hand, the state must help keep alive a propertyless population which has no other means of survival when work is unavailable, maintaining a 'reserve army' of workers through the inevitable cyclical declines in the demand for labor. On the other hand, the state must ensure that escape routes are closed and that means of survival other than wage labor for capital are not so readily available as to liberate the propertyless from the compulsion to sell their labor power when they are needed by capital.[87] This balance may be accomplished, in part, through the promotion or curtailment of labor mobility. Wood explains: "Although the movement of labor across national boundaries has been severely restricted, controlling labor's mobility need not mean

---

84 Steger, *Globalism*, 11.
85 Steger, *Globalism*, 18.
86 Linda Basch, Nina Glick Schiller, and Cristina Szanton Blanc, *Nations Unbound: Transnational Projects, Postcolonial Predicaments, and Deterritorialized Nation-States* (Amsterdam: Gordon and Breach, 1994), 7.
87 Wood, *Empire of Capital*, 18.

keeping workers immobile. It may mean getting them to move to where capital most needs them."[88]

Although capital is generally considered to be more mobile than labor, it is important to recognize that labor also circulates. Saskia Sassen, for example, identifies four systems in which labor importation has played a significant role in the constitution of the labor supply needed for capital accumulation. In the first type, labor was imported into sites of capitalist accumulation that occurred in less 'developed' areas, although the capital was ultimately transferred to 'developed' countries. Examples of this association of labor and capital include the British importation of South Asian workers into the British colonies of present-day Malaysia, and the importation of Japanese workers into the United States-owned plantations of Hawaii. The second form entails the import of labor in connection with capitalist expansion into lesser-developed regions, such as the large-scale movements of laborers to the United States during the nineteenth century. In a third type, labor imports are utilized to replace (or displace) labor in developed countries. This is seen in the practice of using foreign workers to discipline local workers. Lastly, labor importation is used to facilitate the accumulation of capital in developed countries, such as the current importation of workers into Japan.[89]

Contemporary transnational labor migration is intimately related to the wide-scale corporate restructuring (or downsizing) that has been occurring since the early 1970s and, by extension, the growth of contingent (temporary) work. As Robert Parker explains, the transition to an increasingly contingent workforce has provided US businesses with greater flexibility and other benefits but implies several negative effects for workers, including lower wages and the loss of such benefits as health care protection, vacations, pension, and retirement benefits.[90] Guided by a neoliberal philosophy, employers have been aggressively avoiding any long-term commitment to a majority of their workers; employers are not only pursuing a docile workforce willing to work for low wages and few, if any benefits, they are also seeking the flexibility to retain workers only for the briefest periods of time when their labor is required.[91] As discussed by Singer, it is this flexibility that permits private military firms to capitalize on international conflicts through the use of hired labor. Wars become labor markets.

The global space economy has thus been transformed, resulting in a globalized labor market. As expressed in this book, the global labor market consists of a system of integrated local labor markets, connected by international transportation and communication networks. A host of countries, including the Philippines, Pakistan, India, Indonesia, and Sri Lanka, have capitalized on the *global* demand for temporary

---

88 Wood, *Empire of Capital*, 19.

89 Saskia Sassen, *The Mobility of Labour and Capital: A Study of International Investment and Labour Flow* (Cambridge: Cambridge University Press, 1988), 31-32.

90 Robert E. Parker, *Flesh Peddlers and Warm Bodies: The Temporary Help Industry and its Workers* (New Brunswick, NJ: Rutgers University Press, 1994), 2, 138.

91 Parker, *Flesh Peddlers*, 2-3.

workers. Overseas employment programs are utilized by many lower- and middle-income Asian states as a component of national development strategies. Three basic objectives are sought: a reduction in unemployment and underemployment; an increase in human capital as migrant workers return with skills learned abroad; and an increase in foreign revenues through remittances of migrant workers.

Under neoliberalism, the state now aligns with corporate power, transnational corporations, and the forces of militarization. Within the global labor market, a similar privatization revolution has occurred. Private recruitment firms provide the functional integration between internationally dispersed local labor markets. In particular, firms participating in the global labor market are associated with newer and diverse forms of flexible and transnational employment. Truly, the simultaneous globalization and fragmentation of local labor markets, along with technological innovations, have led to a proliferation of diverse employment relations within and not simply across states. As indicated above, the past three decades have witnessed a search by firms for greater flexibility in how work is organized and labor is deployed. Such strategies include job rotation, flexible job arrangements, and the increased use of part-time workers. Transnational contract labor migration provides just one alternative for firms and industries in their adjustment to cyclical and long-term structural changes. Current systems of government-sponsored overseas contract employment, whether originating in the Philippines, Sri Lanka, or any other origin, exhibit similar processes. Potential workers are recruited either by state agencies, private firms, or some combination of the two, and are deployed on temporary contracts, either of six-month or two-year duration. Consequently, through the hiring of contract labor, employers are able to derive benefits from the employment of a temporary, exploitable work-force. This represents a system of labor flexibility on a global scale. In an era of economic neoliberalism, deregulation, and privatization, the promotion of more efficient ways for the unobstructed movement of labor dominates public policy and intellectual discourse in labor-exporting countries such as the Philippines.[92]

Security

Transnational labor migration has been utilized by various firms and governments as a means to facilitate capital. However, these flows also play on the fears on an increasingly globalized world. Indeed, the tragic events of 11 September 2001 generated widespread concern over transnational migration as a security threat. Such increased attention stems from the profiles of the nineteen terrorists involved in the attacks: All were foreign-born, most from Saudi Arabia and Egypt, but residing in the United States on temporary visas. As John Tirman explains, "After the September 11 attacks, the culprits of American vulnerability were widely identified as porous

---

92 Christine B.N. Chin, *In Service and Servitude: Female Foreign Domestic Workers and the Malaysian 'Modernity' Project* (New York: Columbia University Press, 1998), 93.

borders, generous entry policies, violations of the terms of entry, and the entry of immigrants from the Middle East more generally."[93]

As details of the attacks, and of the attackers, began to emerge, it was clear to many that terrorism as a security issue was transnational in scope, dependent on the migration of people, weapons, information, and money.[94] Within days, if not weeks, President Bush and members of Congress hurried into existence a series of directives and acts which targeted immigrants. On October 21, 2001 Bush issued Homeland Security Presidential Directive 2, purported to 'combat' terrorism through immigration policies. Days later Congress enacted the USA PATRIOT Act of 2001 which enlarged the scope of the government to detain and prosecute aliens.

Authorities have responded in kind. Louise Cainkar documents that "of thirty-seven known US government security initiatives implemented since the September 11 attacks, twenty-five either explicitly or implicitly target Arabs and Muslims in the United States."[95] Gary Gerstle elaborates that

> Several months after September 11, 2001, the [US] government asked five thousand men from Middle Eastern and Muslim countries to 'volunteer' for interviews with immigration officials; some of these interviews have triggered deportations. About the same time, the Immigration and Naturalization Service (INS) ordered public and private universities to provide it with information about their Middle Eastern and Muslim students. Hundreds, perhaps thousands, of university students from Middle Eastern countries have already dropped out of school and gone home, and applications from prospective new students have plummeted. In February 2002, the INS began registering and fingerprinting 44,000 immigrants from specified Arab and Islamic countries. A federal noose has tightened around Muslim and Arab immigration, giving the government the ability to choke it off altogether.[96]

This of course is a long-standing preoccupation in the United States. It is often said, usually with pride, that America is a land of immigrants. But immigration has always been selective, and generally regulated to facilitate capital accumulation. The first US federal regulation of immigration, the 1875 Page Law, targeted Chinese workers; subsequent acts were directed against Japanese and Filipino immigrants. Over the years, ideas surrounding the notion of 'acceptable' immigrants began to crystalize, with exclusion acts directed against anarchists, prostitutes, persons likely to become public charges, and other socially undesirable peoples.

---

93 John Tirman, "Introduction: The movement of people and the security of states," In *The Maze of Fear: Security and Migration after 9/11*, edited by John Tirman (New York: The New Press, 2004), 2.

94 Tirman, "Introduction," 2.

95 Louise Cainkar, "The impact of the September 11 attacks on Arab and Muslim communities in the United States," in *The Maze of Fear: Security and Migration after 9/11*, edited by John Tirman (New York: The New Press, 2004), 215.

96 Gary Gerstle, "The immigrant as threat to American security: a historical perspective," In *The Maze of Fear: Security and Migration after 9/11*, edited by John Tirman (New York: The New Press, 2004), 107.

Events following the 2001 attacks were a continuation and a departure of earlier discriminatory acts. Immigrants were singled out as potential criminals—a fifth column lying dormant, waiting to strike at the heart of the United States. What is distinctive about contemporary migration-security concerns lies in the transnational and global character of modern terrorist organizations (at least, as defined by the US State Department). John Gray argues that "As capital has gone global, so has crime." He continues that many "terrorist organizations rely for some of their funding on crime, particularly the trade in illegal drugs. With globalization, they are able to move the funds they acquire from these sources freely around the world."[97] The neoliberal tendencies, including the roll-back of government interference, has therefore benefitted not only multinational corporations in their quest for capital accumulation on a global scale; it has also benefitted those organizations committed to overthrowing Western expansion. Gray finds, for example, that the freedom of capital flows from political control—which benefitted American and other investors—created a vast pool of offshore wealth in which the funds of terrorist organizations can vanish without a trace.[98]

Russell Howard identifies a particularly salient geopolitical implication of transnational terrorism. These groups "do not answer completely to any government; they operate across national borders; and have access to funding and advanced technology. Such groups are not bound by the same constraints or motivated by the same goals as nation-states. And, unlike state-sponsored groups, religious extremists such as Al Qaeda are not susceptible to traditional diplomacy or military deterrence. There is no state with which to negotiate or to retaliate against."[99] In agreement, Alan Duport argues that the attacks on the World Trade Center and the Pentagon underline the growing power of non-state actors to challenge the traditional monopoly over organized violence held by the state. He maintains that "Transnational forces are recasting notions of power and sovereignty away from their traditional rootedness in the territorially bounded state" and it is "this detachment from territory that distinguishes economic and political activity in the twenty-first century."[100] Iain Boal and his colleagues concur, noting that the attacks are also symptomatic of a "new structural feature of the international state system: that the historical monopoly of the means of destruction by the state is now at risk."[101] They explain that this feature has many causes:

---

97 John Gray, *Al Qaeda and What it Means to be Modern* (New York: The New Press, 2003), 74.

98 Gray, *Al Qaeda*, 74.

99 Russell D. Howard, "Understanding al Qaeda's application of the new terrorism—the key to victory in the current campaign," in *Terrorism and Counterterrorism: Understanding the New Security Environment*, edited by Russell D. Howard and Reid L. Sawyer (Guilford, CT: McGraw-Hill, 2004), 75-85; at 78.

100 Alan Dupont, "Transnational violence in the Asia-Pacific: an overview of current trends," in *Terrorism and Violence in Southeast Asia: Transnational Challenges to States and Regional Stability*, edited by Paul J. Smith (Armonk, NY: M.E. Sharpe, 2005), 3-18; at 3-4.

101 Boal et al., *Afflicted Powers*, 31.

Technological advance is one of them. The rise of a worldwide secondary market in arms—partly the result of the chaos attending the end of the Cold War, partly a natural product of the neo-liberal commodification of the globe—is another. Likewise the contracting-out of more and more military services to a shady corporate world, again something that neo-liberalism began by warmly recommending to its client nations. The permeability of borders obviously matters, and has become another major item in the new paranoia. But that fact is linked to a deeper and more pervasive reality, which again is a product of the 'globalization' these same states are committed to—and on which their bloated home economies depend.[102]

Historically, the United States has viewed terrorism as an international force, one that is deliberately directed by governments. International terrorism is viewed from a state-centric perspective, with terrorists being 'state-sponsored.' Consequently, the US State Department is required to report annually to Congress on the patterns of global terrorism and to list the states considered sponsors of terrorism. Trade sanctions are then imposed on these states. Prior to September 11, 2001, seven states were included on the list: Cuba, Iran, Iraq, Libya, North Korea, Sudan, and Syria.[103]

Such an approach, however, is not adequate to explain current trends. More recently, as Howard writes, many intelligence professionals, military operators, pundits, and academics have emphasized the existence of a 'new' terrorism, one that is transnational, borderless, and prosecuted by non-state actors. According to Howard, transnational terrorism is more violent. These terrorists, moreover, are transnational and operate globally; their targets are less individual states and more directed toward ideological systems (e.g., secularism). Transnational terrorists, likewise, are better financed, deriving income from both legal and illegal sources. Lastly, they are composed of networked, cellular structures that are difficult to penetrate.[104]

Al Qaeda is in many respects the exemplar of modern terrorist organizations. Born on the battlefields of Afghanistan following the Soviet Union's invasion of that country, Al Qaeda emerged as a framework "for the sole purpose of creating societies founded on the strictest Islamist principles." Rohan Gunaratna describes Al Qaeda as a "secret, almost virtual, organization." It has, despite claims to the contrary, attempted to remain outside of public scrutiny. Consequently, it assumes other names and identities, such as the World Islamic Front for the Jihad Against the Jews and the Crusaders.[105]

Abdullah Azzam, a Palestinian-Jordanian and mentor of Osama bin Laden, is considered the ideological father of Al Qaeda. In 1987 Azzam defined the group's

---

102 Boal et al., *Afflicted Powers*, 31-32.

103 Louise Richardson, "Global rebels: terrorist organizations as trans-national actors," in *Terrorism and Counterterrorism: Understanding the New Security Environment*, edited by Russell D. Howard and Reid L. Sawyer (Guilford, CT: McGraw-Hill, 2004), 67-73; at 67-68.

104 Howard, "Understanding al Qaeda's application,"at 75-76.

105 Rohan Gunaratna, *Inside Al Qaeda: Global Network of Terror* (New York: Berkley Books, 2003), 4.

composition, aims, and purpose, identifying Al Qaeda as the vanguard of a pan-Islamic ideology. Consequently, with the end of the anti-Soviet Afghan *jihad*, large numbers of Arab and Asian *muhahidin* returned to their home countries and joined opposition political parties, religious bodies and other groups, campaigning against corrupt regimes.[106]

Al Qaeda's global network consists of permanent or independently operating semi-permanent cells of trained militants that have been established in more than seventy-six countries. Military training camps have been established in Sudan, Yemen, Chechnya, Tajikistan, Somalia, and the Philippines.[107] According to Howard, since 2001 more than 3,300 Al Qaeda operatives, hailing from 47 different countries, have been arrested in 97 countries.[108] Al Qaeda also has developed a series of transnational ties with other organizations. Howard, for example, cites evidence linking Al Qaeda with the Syrian- and Iranian-backed Lebanese group Hezbollah; Al Qaeda is also linked with various groups in Southeast Asia, including the Philippines-based Abu Sayyaff Group and Jemaah Islamiya (JI).[109]

Al Qaeda is global in its activities, although its strategic objectives have always been more concrete and limited.[110] This is perhaps best illustrated in a lengthy 1996 document—"Declaration of War Against the Americans Occupying the Land of the Two Holy Places"—disseminated by Bin Laden. This document contains, in detail, both his objectives and his planned means of achieving them. James Robbins summarizes Bin Laden's four principle strategic goals. The first is to expel the United States military from the Arabian peninsula. As discussed in Chapter 2, the United States has maintained a military presence—in intent if not in actual military forces—throughout much of the latter twentieth century. Bin Laden believes it to be a sacrilege to have the 'infidel armies of the American crusaders' so close to the holy places of Islam. A second goal is the overthrow of the corrupt Muslim (and particularly Arab) regimes, and the restoration of the Caliphate. Bin Laden thus supports the formation of a pure Islamic state, guided by the *Sharia*. To accomplish this, however, it is necessary to remove the only foreign superpower—the United States—from the region. A third goal is the destruction of the state of Israel and the creation of a Palestinian homeland, while a fourth goal is to punish the United States for its global acts of aggression against Muslims.[111]

Ironically, Al Qaeda's thoroughly 'modern' program resonates with contemporary anti-globalization campaigns. Gray explains that Al Qaeda is essentially a modern

---

106 Gunaratna, *Inside Al Qaeda*, 6.

107 Gunaratna, *Inside Al Qaeda*, 7.

108 Howard, "Understanding al Qaeda's application," 77.

109 Howard, "Understanding al Qaeda's application," 77; see also Paul J. Smith (editor), *Terrorism and Violence in Southeast Asia: Transnational Challenges to States and Regional Stability* (Armonk, NY: M.E. Sharpe, 2005).

110 Gray, *Al Qaeda*, 75.

111 James S. Robbins, "Bin Laden's War," in *Terrorism and Counterterrorism: Understanding the New Security Environment*, edited by Russell D. Howard and Reid L. Sawyer (Guilford, CT: McGraw-Hill, 2004), 392-404; at 392-93.

organization in that it not only uses satellite phones, laptop computers, and encrypted websites, but that it also understands that "twenty-first century wars are spectacular encounters in which the dissemination of media images is a core strategy." He elaborates that Al Qaeda is also very modern in its organization: it "resembles less the centralized command structures of twentieth-century revolutionary parties than the cellular structures of drug cartels and the flattened networks of virtual business operations."[112] Al Qaeda, essentially, is a multi-national corporation, one that is not guided by territorial jurisdiction. Indeed, in a double irony, Al Qaeda's 'resistance' to globalization is in fact supported by many of the same globalizing tendencies and transnational practices it opposes.

The Path Ahead

Occupied Iraq is neither a sovereign state nor a colony. 'Authority' and, supposedly, sovereignty, were transferred to the Iraqi government on June 28, 2004. In January 2005, elections were held. And yet an American military presence remains firmly entrenched. The statements of President George Bush and Secretary of Defense Donald Rumsfeld, in particular, give no indication that this state of affairs will soon change. America, it is claimed, is 'in it' for the long run. Occupied Iraq is, and following the work of Giorgio Agamben, a space of the exception. It is a zone not of exclusion, but of abandonment, a space where lines between right and wrong, legal and illegal are blurred. Occupied Iraq has been produced—for this was not a natural or inevitable occurrence—in the twenty-first century, as a de-humanized place.

How are we to interpret the occupation of Iraq? What is the place of hostages in Occupied Iraq? The broaching of these questions—and the search for answers— is not to discount the suffering, nor to diminish the loss of life in Occupied Iraq. Rather, and following Orwell, my concern lies with understanding the meanings of the invasion and occupation with an eye toward future conflicts. Orwell was troubled by his observations of human suffering, of the deplorable conditions, the places in which the poor, the marginalized, the dispossessed lived and died. Orwell explained that his writings were impelled by a "Desire to push the world in a certain direction, to alter other people's idea of the kind of society that they should strive after."[113] He continued that "Every line of serious work that I have written since 1936 has been written, directly or indirectly, against totalitarianism and for democratic socialism, as I understand it." To Jeremy Meyers, Orwell's socialism "was not based on political and economic principles, but on liberal and humane beliefs."[114] My purpose is to de-center the political subjugation of hostages toward a progressive politics. Postcolonialism, as Robert Young explains, "seeks to intervene." It seeks "to change the way people think, the way they behave, to produce a more just and equitable

---

112 Gray, *Al Qaeda*, 76.

113 George Orwell, *A Collection of Essays* (New York: Harcourt, 1981), 312-3.

114 Jeffrey Meyers, *Orwell: Wintry Conscience of a Generation* (New York: W.W. Norton & Co., 2000), 90.

relation between the different peoples of the world."[115] Orwell likewise used his writings to understand the place of humanity within de-humanizing, totalitarian worlds, regardless of political, economic, or religious orientation. We can do no less.

The war in Iraq, I maintain, was not (and still is not) reducible to oil. Hostages were not (and are not) sacrificed solely to lubricate the economies of the United States and its 'willing' coalition members. Indeed, the Iraqi war constitutes "a radical, punitive, 'extra-economic' restructuring of the conditions necessary for expanded profitability—paving the way, in short, for new rounds of American-led dispossession and capital accumulation." The war "was a hyper-nationalist neo-liberal *putsch*, made in the name of globalization and free-market democracy."[116]

Workers, warriors, and hostages. All have been, and continue to be, exploited in Occupied Iraq situated within a larger globalized struggle of profits. This is the foundation upon which *The Business of War* is based.

---

115 Robert J.C. Young, *Post-Colonialism: A Very Short Introduction* (Oxford: Oxford University Press, 2003), 7.

116 Boal et al., *Afflicted Powers*, 72.

Chapter 2

A War of Neoliberalism

On 20 March 2003, at 10:16 pm (eastern standard time), President George W. Bush addressed the American public and the world as a whole. In a brief four-minute televised appearance, Bush explained that "American and coalition forces [were] in the early stages of military operations to disarm Iraq, to free its people and to defend the world from grave danger." Bush indicated that the war would be a "broad and concerted" campaign. However, he also warned that "helping Iraqis achieve a united, stable and free country" would require a "sustained commitment." Just how long the commitment would last was left unsaid. Rather, Bush portrayed the war as inevitable, but that the response would be overwhelming in its force. He said that "Now that conflict has come, the only way to limit its duration is to apply decisive force."[1]

The invasion was further justified by the administration with claims that Saddam Hussein and his Iraqi regime possessed weapons of mass destruction and were capable of launching these devices 'imminently.' Bush explained that "The people of the United States and our friends and allies will not live at the mercy of an outlaw regime that threatens the peace with weapons of mass murder." Missing from his announcement, curiously, was any reference to terrorism or Al Qaeda. Rather, Bush explained that "We come to Iraq with respect for its citizens ... We have no ambition in Iraq, except to remove a threat and restore control of that country to its own people."

The war was quick and decisive. By 6 April Baghdad had fallen. On 14 April Pentagon officials indicated that major combat operations were over. And on 1 May Bush declared an end to combat operations. In a televised address from the aircraft carrier USS Abraham Lincoln (off the coast of San Diego, California), Bush declared that "In the battle of Iraq, the United States and our allies have prevailed. And now our coalition is engaged in securing and reconstructing that country."[2] Discursively, Bush framed the military operation as a battle—one component of the larger War on Terror. The re-positioning is extremely important, as it re-directs attention away from the actual practices associated with the occupation of Iraq. Portrayed as a 'battle' in a larger war, the invasion of Iraq was—according to the Bush administration— intimately tied to a campaign against terror. This enabled the Bush administration

---

1   Office of the Press Secretary, "President Bush Addresses the Nation," March 19, 2003, www.whitehouse.gov/news/releases/2003/03/2030219.17.html (April 12, 2004).

2   "A Crucial Advance in the Campaign Against Terror," *The Guardian*, May 1, 2003, www.guardian.co.uk/print/0.3858.4660759-103550.00.html (October 15, 2003).

to play on the fears of porous national boundaries, immigration, and, consequently, future attacks. The American public is more likely, so the argument goes, to sacrifice personal liberties in the name of security.

In George Orwell's nightmarish *Nineteen Eighty-Four*, wars were waged for the purpose of conducting future wars. The aim of warfare was not to win, for wars had become unwinnable, but rather to use the products of the war machine—the military-industrial complex—without raising the general standard of living. War had become a business. War was essential to facilitate the continued production, circulation, and consumption of war-related materials. Orwell wrote: "The essential act of war is destruction, not necessarily of human lives, but of the products of human labor. War is a way of shattering to pieces, or pouring into the stratosphere, or sinking in the depths of the sea, materials which might otherwise be used to make the masses too comfortable, and hence, in the long run, too intelligent."[3]

As discussed in Chapter 1, capitalism is driven by certain systemic imperatives, namely the imperatives of competition, profit-maximization, and profit-accumulation.[4] Indeed, as David Harvey explains, accumulation is the engine which powers growth under the capitalist mode of production.[5] Capitalism, however, is also inherently and inevitably crisis-prone. Termed variously as a *crisis of over-accumulation, over-production,* or *over-capacity,* crises are marked by chronic unemployment and underemployment, capital surpluses and lack of investment opportunities, falling rates of profit, and a lack of effective demand in the market.[6] Confronted with crises, states may intervene in any number of ways, including a process which Harvey terms a 'spatial fix.' Through this process, excess capital and labor may be absorbed through geographic expansion. Capitalism is 'spread' to new territories through trade, foreign investment, and the outsourcing of employment.

The 'War on Terror' is a never-ending war, born of a militant neoliberal capitalist imperative. It constitutes a discursive construct, one that enables the Bush administration to justify and enact foreign policies that would otherwise be opposed, such as the illegal invasion and occupation of Iraq. Orwell 'warned' in his writings that "it is often necessary for a member of [the administration] to know that this or that item of war news is untruthful, and he may often be aware that the entire war is spurious and is either not happening or is being waged for purposes quite other than the declared ones; but such knowledge is easily neutralized by the technique of doublethink."[7] Members of the Bush administration routinely and blatantly engaged in the technique of doublethink. Questionable and unfounded connections were alleged, all geared toward the manufacture of mass hysteria and collective support for a pre-determined invasion and occupation of Iraq.

---

3   George Orwell, *Nineteen Eighty-Four* (New York: Plume, 1983 [1949]), 169-71.

4   Ellen Meiksins Wood, *Empire of Capital* (London: Verso, 2003), 10.

5   David Harvey, *Spaces of Capital* (New York: Routledge, 2001), 237.

6   Walden Bello, *Dilemmas of Domination: The Unmaking of the American Empire* (New York: Henry Holt and Company, 2005), 4; Harvey, *Spaces of Capital*, 240.

7   Orwell, *Nineteen Eighty-Four*, 171.

The war is real. It is measured in the lives of citizens and soldiers, workers and warriors, men, women, and children. The war, however, was *not* waged to ensure the security of the United States; it was not waged to prevent Saddam Hussein from *using* weapons of mass destruction; and it was not waged to prevent a future Al Qaeda attack. Years in the making, the war and subsequent occupation of Iraq was waged for material gain that—ultimately—would benefit only a select few corporate CEOs, financiers, and bankers.

A month before the invasion of Iraq, Nicholas Lemann wondered: "Has a war ever been as elaborately justified in advance as the coming war with Iraq? Because this war is not being undertaken in response to a single shattering event ... and because the possibility of military action against Saddam Hussein has been Washington's main preoccupation for the better part of a year, the case for war has grown so large and variegated that its very multiplicity has become a part of the case against it."[8] Iain Boal and his colleagues write of the "sheer ponderousness and hypocrisy of the pre-war build-up—and its constant *visibility*, the repeated set-pieces of lying and bullying to the world at large."[9]

Is the war on Iraq exceptional? Or does it reflect a continuation of deeper, long-term trajectories of United States foreign policy. Phrased differently, does the Bush Doctrine of preemption mark a break in foreign policy? Michel Foucault warned against the search for origins and, in the process, questioned the initial quest or 'discovery' of continuities and discontinuties. Through his writings, Foucault cautioned us to temporarily suspend certain assumptions. We must question, for example, the *principle of coherence*, the idea that there exists ready-made syntheses that are normally accepted before any examination. He suggested that it should never be possible to assign the irruption of a real event; rather, "one is led inevitably, through the naïveté of chronologies, towards an ever-receding point that is never itself present in any history." Consequently, "all beginnings can never be more than recommencements or occultation."[10] In this chapter I argue that the unilateralism evinced in the present Bush administration does not constitute a radical rupture or discontinuity from existing US foreign policy. Rather, the actions of the Bush administration reflect a continuity with earlier administration, namely American imperial expansion through military intervention. It was this historical geography of empire-building that set the stage for the invasion and occupation of Iraq.

---

8   Nicholas Lemann, "After Iraq," *The New Yorker*, February 17, 2003, www.newyorker. com/printables/fact/030217fa_fact (August 30, 2005).

9   Iain Boal, T.J. Clark, Joseph Matthews, and Michael Watts, *Afflicted Powers: Capital and Spectacle in a New Age of War* (New York: Verso, 2005), 4.

10   Michel Foucault, *The Archeaology of Knowledge and the Discourse on Language*, trans. A.M. Sheridan Smith (New York: Pantheon Books, 1972), 22.

Imperial America

Military excursions have served as the primary strategic element in a history of
relentless imperial expansion—from the Monroe Doctrine through the Cold War
to the present.[11] Initially, military force was used to remove European claims to the
'New World.' Having defeated the British in the War of 1812, the United States began
to utilize its newly enhanced and professionalized armies to expand the geography
of American power. Between 1810 and 1813, the United States acquired most of the
western Florida territory between New Orleans and Mobile. By 1819 General Andrew
Jackson forced Spain to cede its remaining possessions to the United States. The
primary motivation for early expansion across the North American continent lay in
the search for markets to service an increasingly commercial agrarian economy.[12]

American expansion was not confined to the shores of the North American
continent. In 1823 President James Monroe announced that the American continent
was effectively 'off-limits' for European colonization. Crucially, this doctrine
threatened preemptive military action in response to any European challenge to its
planned hegemony in the Western Hemisphere.[13] US corporations, rubber merchants,
gun manufacturers, railroad companies, and oil prospectors, would later find ample
opportunities for business investments in the territories 'protected' by the Monroe
Doctrine.[14]

In the 1830s and 1840s agricultural and commercial needs fed a desire to acquire
the western territories of North America. In 1803 the United States purchased the
Louisiana territories, a vast stretch of land extending westward from the Mississippi
River to the Rockies and northward to Canada. Of principle importance was that
the purchase included New Orleans, a city that already served as a key outlet for
the productions of the United States. Subsequently, under the presidency of James
Polk, the US oversaw the violent annexation of Texas from Mexico in 1845 and
the possession of California following a war against Mexico in 1846. Indeed, as
David Slater writes, it was on the eve of the US-Mexican War and in the wake
of the annexation of Texas that notions of 'Manifest Destiny' came to dominate
the ideological underpinnings of American foreign policy. A term coined by John
L. O'Sullivan, the editor of the *Democratic Review*, 'manifest destiny' spoke of a
boundless future for America. It was O'Sullivan who, in 1845, wrote of 'our manifest
destiny to overspread the continent allotted by Providence for the free development of
our yearly multiplying millions." Slater argues that the doctrine of 'manifest destiny'
embraced a belief in American Anglo-Saxon superiority, and it was deployed to
justify war and the appropriation of approximately 50 percent of Mexico's original

---

11   Boal et al., *Afflicted Powers*, 80; see also 82-93.

12   Boal et al., *Afflicted Powers*, 82.

13   At the time other colonial powers, such as the United Kingdom and France, well
understood the inability of the United States to enforce such as declaration. Nevertheless,
Monroe established an arrogance that has remained central to American foreign policy ever
since.

14   Neil Smith, *The Endgame of Globalization* (New York: Routledge, 2005), 49-50.

territory. At the end of the war in 1848, President Polk explained disingenuously that the territories lost by Mexico would have continued to remain of "little value to her or to any nation, while as part of our Union they will be productive of vast benefits to the United States, to the commercial world, and the general interests of mankind."[15]

This American demonstration of willingness to wage war, according to Boal and his colleagues, would later impel Britain to cede the Oregon territory to the US, and compelled the French and Spanish to back away from the Yucatán.[16] Polk, in his 1848 State of the Union address, would described the benefits of territorial expansion:

> No section of our country is more interested or will be more benefitted than the commercial, navigating, and manufacturing interests of the Eastern States. Our planting and farming interests in every part of the Union will be greatly benefitted by it. As our commerce and navigation are enlarged and extended, our exports of agricultural products and of manufactures will be increased, and in the new markets thus opened they can not fail to command remunerating and profitable prices.[17]

Territorial expansion continued throughout the nineteenth century. At the instigation of William Seward, the United States purchased the Alaskan territories in 1867. Reflecting a pragmatic viewpoint, Seward believed that the United States needed massive development of its infrastructure, better internal social cohesion, and vigorous action to secure worldwide shipping stations and open corridors of trade.[18] The ideas of Seward were continued during the administrations of William McKinley and Theodore Roosevelt. In 1898 the United States acquired the Philippines, Guam, and Puerto Rico as part of the spoils of the Spanish-American War. Cuba, in the course of war, was ostensibly 'liberated', but the conditions of the Platt Amendment (1901) ensured that America could, and would, intervene if need be. Indeed, American forces were dispatched to Cuba four times by 1920.

America's decision to colonize the Philippines was in part shaped by a prolonged evolution of US policy in both the Caribbean and the Pacific. American colonial expansion in the Pacific was gradual but steady, marked by Commodore Matthew Perry's 'opening' of Japan and the possession of dozens of remote islands in the Pacific, including the Hawaiian Islands and parts of Samoa. Many of these territorial acquisitions were the result of an enlarged maritime presence of the United States, especially as they were related with both the whaling industry and the strengthening US Navy. American intervention in the Caribbean was no less extensive. Under Theodore Roosevelt's corollary to the Monroe Doctrine, the United States assumed

---

15 David Slater, *Geopolitics and the Post-Colonial: Rethinking North-South Relations* (Malden, MA: Blackwell, 2004), 36.

16 Boal et al., *Afflicted Powers*, 83.

17 Leonard Dinnerstein, Roger L. Nichols, and David M. Reimers, *Natives and Strangers: A Multicultural History of Americans* (New York: Oxford University Press, 1996), 86.

18 Anders Stephanson, *Manifest Destiny: American Expansion and the Empire of Right* (New York: Hill and Wang, 1995), 62.

the role of international 'policeman' of the Western Hemisphere. Between 1904 and 1934 the United States sent eight expeditionary forces to Latin America, took over customs collections twice, and conducted five military occupations.[19]

The American colonization of the Philippines and other Pacific and Caribbean possessions was influenced also by a broader international environment of imperial capitalist expansion. During the latter part of the nineteenth century the industrializing powers of northwest Europe competed for colonies throughout Africa and Asia. Anders Stephanson notes that between 1875 and 1914, one quarter of the world was claimed as colonies; Britain alone added 4 million square miles.[20] Within the United States, imperialists such as Alfred Thayer Mahan and Henry Cabot Lodge advocated for a strong American presence abroad. Extending the idea of America's 'manifest destiny,' the imperialists argued that only as a world power could the United States trade, prosper, and protect itself against potential enemies.[21]

From the late nineteenth century onwards, the US gradually learned to mask the explicitness of territorial gains and occupations under the mask of a space-less universalization of its own values.[22] A recurrent theme of American politics, for example, has been that of America as the beacon of freedom. President Andrew Jackson, for example, referred to US expansion as "extending the area of freedom" while President William McKinley declared that "We intervene [in the Philippines] not for conquest. We intervene for humanity's sake [and to] earn the praises of every lover of freedom the world over." [23] From politicians to pundits to poets, America has been constructed as the Promised Land, its values and beliefs held as universal. As Clyde Prestowitz writes, America has been defined not by a set of ideologies, but rather as an ideology itself. The key principles of Americanism include the ambiguous—but no less powerful—concepts of liberty, equality, individualism, populism, and limited government.[24]

America and Americans have also been constructed as the exception. As David Howard-Pitney writes, "No belief has been more central to American civil religion that the idea that Americans are in some important sense a chosen people with a historic mission to save and remake the world."[25] This remaking would include the export of American values and beliefs, including democracy, liberty, and freedom. Indeed, the idea of freedom, as developed in the United States, has always entailed a sense of morality and responsibility. This is derived, in part, from the seventeenth-

---

19 Ivo H. Daalder and James M. Lindsay, *America Unbound: The Bush Revolution in Foreign Policy* (Washington, D.C.: Brookings Institution Press, 2003), 5.

20 Stephanson, *Manifest Destiny*, 72.

21 Stanley Karnow, *In Our Image: America's Empire in the Philippines* (New York: Random House, 1989), 10.

22 David Harvey, *The New Imperialism* (Oxford: Oxford University Press, 2003), 47.

23 Clyde Prestowitz, *Rogue Nation: American Unilateralism and the Failure of Good Intentions* (New York: Basic Books, 2003), 30-33.

24 Prestowitz, *Rogue Nation*, 36.

25 David Howard-Pitney, *The Afro-American Jeremiad: Appeals for Justice in America* (Philadelphia: Temple University Press, 1990), 6.

century liberalism of John Locke and America's Puritan heritage. As such, American exceptionalism has dovetailed nicely with justifications for territorial and economic expansion.

Arguably, no president has promoted these visions more than Woodrow Wilson. Wilson shared with all his predecessors an unwavering belief in American exceptionalism. However, Wilson differed from previous presidents in that he forwarded a liberal internationalism. He set forth a moral argument for broad American engagement in world affairs.[26] Wilson, for example, declared war against Germany not because Germany endangered American interests, but instead because the United States, from Wilson's perspective, must fight to make the world safe for democracy. It was this commitment to a world in which democracy could flourish that formed the basis of the League of Nations, established with the Treaty of Versailles. Wilson was convinced that the idea of collective security, combined with territorial integrity and political independence would prevent war and remake world politics.[27] The US Senate, however, opposed Wilson's geopolitical vision, largely on the grounds that Wilson's internationalism threatened to entangle the country in foreign alliances.

Hegemony over the Caribbean and a few isolated outposts in the Pacific is one thing. Global hegemony is another. As discussed by Colin Flint and Ghazi-Walid Falah, a hegemonic power is a single state that obtains dominance in the three separate but related economic spheres of production, trade and finance. Moreover, there are crucial geographic processes that underlie the creation and projection of hegemonic power. Flint and Falah maintain that "hegemony is the global diffusion of economic, political and cultural practices originating from the activities of one nation-state."[28]

Although the roots of American hegemony can be traced back to the founding of the state, it was not until the mid-twentieth century that the United States was able to effectively shape the global order. The end of the Second World War saw the United States emerge as the most powerful state on earth. It was relatively untouched by war, and its enormous industrial potential had produced war matériel not only for its own armies and navies but for those of its allies as well.[29] As summarized by Daalder and Lindsay, the foreign policy questions Americans faced at the end of World War II had little to do with what the United States *could* do abroad; the immediate questions dealt with what the United States *should* do abroad.[30]

Reflecting a continuation with earlier periods, American officials in the post-war years claimed to act in universal interests. American politico-military influence

---

26 Daalder and Lindsay, *America Unbound*, 6.

27 Daalder and Lindsay, *America Unbound*, 6-7.

28 Colin Flint and Ghazi-Walid Falah, "How the United States Justified it War on Terrorism: Prime Morality and the Construction of 'Just War'", *Third World Quarterly* 25(2004): 1379-1399; at 1389.

29 Amos A. Jordan, William J. Taylor, Jr., and Michael J. Mazaar, *American National Security*, 5th ed. (Baltimore: Johns Hopkins University Press, 1999), 66.

30 Daalder and Lindsay, *America Unbound*, 8.

would be used to spread democracy, freedom and liberty to all corners of the earth. Elite elements within the United States government and private sector, accordingly, outlined post-war settlement plan that would ostensibly guarantee peace, economic growth and stability. US policy-makers discounted claims of territorial aggrandizement, as American interests were defined as global interests. However, practices more often than not were designed to facilitate US hegemonic ambition.[31]

Strategically, American officials attempted to re-create a global world order that was conducive to American capitalist expansion. Represented as the chief defender of freedom (understood in terms of free markets) and the rights of private property, American presidents provided economic and military protection for pro-US allies and client regimes throughout Europe, Asia, and the Middle East.[32] In turn, the US intervened, both covertly and overtly, in the foreign affairs of other governments. The objective was simple: "the US sought to construct an open international order for trade and economic development and rapid capital accumulation along capitalistic lines."[33]

Increasingly, the Middle East was viewed as a vital strategic location. Since the 1920s, Western powers, including Britain, the United States, France, and the Netherlands, pursued economic interests—namely oil—in the region. These endeavors often reflected the dominance of private industries. Oil exploration in Iraq, for example, was pursued largely under the auspices of the colonial-controlled Iraqi Petroleum Company. Although Iraq gained its political independence in 1932, the country remained bound within a neo-colonial relationship, with its economy dominated by the petroleum company.

During the 1940s both Western and Soviet politicians interfered in the state-building processes of the Middle East and neighboring regions. In 1947 the Truman Doctrine, for example, asserted that it was the policy of the United States to support free peoples resisting subjugation in the maintenance of their freedoms. In 1947-1948 the United States also played an influential role in the creation of Israel and, in subsequent years, would remain committed to the new state. According to Amos Jordan and colleagues, the United States supported Israel largely for moral and domestic political reasons. The United States likewise worked to protect and improve the position of American companies developing the oil resources of the Persian Gulf region as well as to preserve overflight and transit rights through the Middle East to areas of greater strategic importance.[34] Truman's blanket commitments, based on an opposition to communism and a concern with Soviet expansion in the eastern Mediterranean and northern Gulf states in the Middle East, would largely define United States foreign policy for the next 45 years.

The hallmark of Truman's foreign policy, according to Daalder and Lindsay, was its blend of power and cooperation. Truman was "willing to exercise America's

---

31 Harvey, *New Imperialism*, 50.
32 Harvey, *New Imperialism*, 51-52.
33 Harvey, *New Imperialism*, 54.
34 Jordan, et al., *American National Security*, 394-95.

great power to remake world affairs, both to serve American interests and to advance American values."[35] Truman did so, however, through a complex web of alliances and institutions. He oversaw the creation, for example, of the North Atlantic Treaty Organization (NATO), the United Nations (UN), the International Monetary Fund (IMF), the World Bank, the General Agreement on Tariffs and Trade (GATT), and the Organization of American States. In creating these institutions, as Daalder and Lindsay point out, Truman established a precedent that the United States had the power to act as it saw fit, but that its right to act should be constrained by international law.[36] This should not hide the fact, however, that the United States was able to unduly exert its influence, and thus promote its investments, through the manipulation of these transnational institutions.

American activities within the Middle East, however, were largely limited. For the most part, the United States sought to escape becoming entangled in local affairs. American policy-makers evaluated specific events in the Middle East in terms of how they affected US interests elsewhere—namely Europe—and reacted to them accordingly.[37]

American commitments in the Middle East were more forcefully expanded during the presidential administration of Eisenhower. This is seen most clearly in the US-supported overthrow of Prime Minister Mohammed Mossadeq in Iran. First elected to the Iranian parliament in 1923, Mossadeq had been prime minister since 1951. It was Mossadeq who, in the early 1950s, pledged to remove the Anglo-Iranian Oil Company (AIOC) from Iran, reclaim the country's vast petroleum reserves, and free Iran from subjection to foreign power. Since the early part of the twentieth century the AIOC had enjoyed a lucrative monopoly on the production and sale of Iranian oil. This wealth largely supported the British empire while preventing Iran from development.[38]

As Prime Minister, Mossadeq nationalized the AIOC. The British, in response, accused Mossadeq of stealing their property. They demanded that Mossadeq be punished by the World Court and the United Nations, and later sent warships to the Persian Gulf. Ultimately, Britain imposed a strict embargo that devastated Iran's economy. Plans were also underway to remove Mossadeq from power.[39] At this point, Britain asked the United States for assistance.

While still in office, Truman, in part, sympathized with the nationalist movement in Iran. According to Stephen Kinzer, Truman had nothing but contempt for old-style imperialists like those who ran the Anglo-Iranian Oil Company. Perhaps most salient, however, was that the US Central Intelligence Agency (CIA) had never overthrown

---

35 Daalder and Lindsay, *America Unbound*, 9.

36 Daalder and Lindsay, *America Unbound*, 10.

37 Jordan, et al., *American National Security*, 395.

38 Stephen Kinzer, *All the Shah's Men: An American Coup and the Roots of Middle East Terror* (New York: John Wiley & Sons, 2003), 2.

39 Kinzer, *All the Shah's Men*, 2-3.

a government, and Truman did not want to set the precedent.[40] Eisenhower, unlike Truman, was not opposed to the idea of overthrowing the Iranian prime minister. His rationale, though, centered less on 'saving' British interests than it did on containing communism. A pro-Western Iran, Eisenhower, hoped, would help promote American presence in the region. To be sure, Iran had immense oil wealth; the country also shared a long border with the Soviet Union and was home to an active Communist party. Eisenhower, along with his Secretary of State John Foster Dulles and incoming CIA director Allen Dulles, were all in agreement that America should put in place a reliably pre-Western prime minister.[41]

In 1953, as part of a US covert operation known as Operation AJAX, the democratically-elected Mossadeq was overthrown. Subsequently, the monopoly of the AIOC was broken as five major American oil companies, along with Royal Dutch Shell and the French Compagnie Française del Pétroles were given licenses to operate in Iran. As Roger Burbach and Jim Tarbell explain, the "coup reshaped the tactics, if not the strategies, of the cold war." They contend that the "constant use of covert interventions combined with the periodic deployment of military forces in effect transformed and transcended the meaning of 'informal' and 'formal' empires." As the "self-proclaimed leader of the 'free world" the United States "acted as an arrogant and self-serving imperial power determined to impose its will on the world."[42] And the results, as Boal and colleagues write, were 'dizzying': the staging of a coup in Guatemala (1954); the landing of Marines in Lebanon (1958); the deployment of 20,000 troops to the Dominican Republic (1965); the overthrow of the civilian government in Indonesia (1965); the installation (1954) and removal of Ngo Dinh Diem in South Vietnam (1963); the defeat of Salvador Allende in Chile (1973); a proxy war in Nicaragua (1970s-1980s); a 'civil war' in El Salvador (1980s); another incursion into Lebanon (1982-1983); and the invasion of Grenada (1983) and Panama (1989).[43]

American activities in the Middle East, while still relatively minor compared to US actions in Asia and Europe, were solidified in 1957 with the Eisenhower Doctrine. Concerned with the possibility of Soviet aggression, the Eisenhower Doctrine pledged the United States to defend the Middle East from any over armed aggressions from any nation controlled by International Communism. These attitudes persisted, and indeed intensified, during the 1960s and the administration of John F. Kennedy. Three changes, in particular, led to an increased US presence in the Middle East. First, several US interests, such as American oil companies, which previously had been marginal to the domestic American political and economic groups, were elevated to the level of important or even vital national interests. Second, the growing Soviet presence and declining British power made it important

---

40 Kinzer, *All the Shah's Men*, 3.

41 Kinzer, *All the Shah's Men*, 4.

42 Roger Burbach and Jim Tarbell, *Imperial Overstretch: George W. Bush and the Hubris of Empire* (New York: Zed Books, 2004), 64-65.

43 Boal et al., *Afflicted Powers*, 88.

that alternative means be found to protect American interests in the region. Third, it was increasingly recognized that various US interests in the Middle East which had been treated separately, such as Israel and Persian Gulf oil, were intertwined.[44]

Iraq increasingly occupied the attention of US policy-makers. Within Iraq, the pro-British constitutional monarchy of Iraq was overthrown in a military coup. The action brought Abdul Karim Kassem to power. In response, the United States deployed troops to the region in an attempt to pressure Iraq into conformity with Western interests. Kassem, however, persisted with his nationalist program. In 1959 he removed Iraq from the Baghdad Pact. Established in 1953, the Baghdad Pact was a NATO-sponsored agreement whose original members included Iraq, Iran, Pakistan, Turkey, and the United Kingdom. The purpose of the pact had been to contain "Nasirism," a socialist-derived form of Arab nationalism expounded by Egyptian president Abd-al-Nasir. These developments would set in place an intensified interest in the country for years to come. Beginning with the ascension of Kassem, Western governments contemplated the removal of the nationalist leader. Throughout this brief reign as premier of the new Iraqi Republic, Kassem suppressed opposition and worked to bring about social reform. He attempted to remain neutral within the Cold War and was bitterly opposed to imperialism. It was his attempt to nationalize oil, however, that was most threatening to Western, and especially American, interests. In early 1963, Kassem announced the formation of a national oil company to exploit the areas he had expropriated from foreign companies in 1961. The territory involved amounted to 99 percent of their concessionary areas. In February 1963 Kassem was toppled in a coup—allegedly supported by Western governments—and was eventually killed.[45]

Set within the debacle of Vietnam, American officials attempted to influence events in the Middle East while not becoming embroiled in a potential quagmire. President Richard Nixon, aware of the very real possibilities of imperial over-stretch, attempted to phase down American commitments in the world. The Nixon Doctrine, as it emerged over time, asserted that the United States would continue to participate in the defense and development of friends and allies, but that the US would not undertake all aspects of defense. The United States, in short, would give first priority to its own interests.[46] In practical terms, the Nixon administration enhanced the military strength of Iran and Saudi Arabia so that they could take greater responsibility for the region's and their own security.[47]

American interventions did not necessarily lead to stability in the Middle East. Nor did American actions lead to pliable, pro-Western governance. Following the

---

44 Jordan, et al., *American National Security*, 396.

45 Geoff Simons, *Iraq: From Sumer to Post-Saddam* (New York: Palgrave MacMillan, 2004), 258-59.

46 John Lewis Gaddis, *Strategies of Containment: A Critical Appraisal of Postwar American National Security Policy* (New York: Oxford University Press, 1982), 298.

47 Michael T. Klare, *Blood and Oil: The Dangers and Consequences of America's Growing Petroleum Dependency* (New York: Metropolitan Books, 2004), 43.

ouster of Kassem, for example, Iraq was beset with five years of internal strife, as various nationalist coalitions attempted to gain control of the country. In the end, it was the Ba'ath (Resurrection) Party that gained power. Originating out of the struggle for national, post-colonial identity in Syria, the Ba'ath Party spread throughout Lebanon, Jordan, and Iraq. Socialist, secular, and modern in its orientation, the Ba'ath Party emerged as a dominant nationalist movement in the Middle East.[48]

It was during the late 1950s and 1960s that Saddam Hussein began to emerge as a powerful political force. In 1959 he had participated in a failed assassination attempt of Kassem; he was subsequently arrested and jailed until 1966. When the Ba'ath Party assumed control of Iraq in 1968, Saddam Hussein became vice president in charge of oil. Over the next decade Saddam increasingly consolidated his power until, in 1979, he staged a 'palace coup' and assumed full dictatorial control of the country.[49] Saddam attempted to become the uncontested leader of the Pan-Arab world, predicated on a nationalist and mostly secular ideology. Economically, he sought to industrialize the country and, to this end, he nationalized the Iraqi Petroleum Company in 1972. Hussein also attempted to play-off the two superpowers. His moves to the Soviet Union, however, contributed to a hardening of US-Iranian relations.

The geopolitical terrain of the Middle East was radically altered in 1979. In February the US-based shah of Iran was deposed, and the subsequent Iranian revolution called for the creation of Islamic states throughout the Middle East. In particular, the Ayatollah Khomeini issued a call to abolish Arab or state nationalism in the interest of Islamic unit and began agitating against pro-US oil sheikdoms in the Gulf. Later that year, Islamic students seized the US embassy in Tehran with Khomeini's blessing and held 52 Americans hostage for 444 days. Concerns were further raised by the Soviet Union's invasion of Afghanistan of that same year.

Michael Klare explains that the fall of the shah spelled the failure of the Nixon Doctrine in the Persian Gulf and forced the American government to conduct a fresh review of security policy in the region.[50] The Carter Doctrine formally underscored the Persian Gulf's centrality to US global power and America's willingness to wage war to maintain its regional dominance. As articulated in his 1980 State of the Union address, President Jimmy Carter asserted that any attempt by any outside force to gain control of the Persian Gulf would be regarded as an assault on the vital interests of the United States.[51]

The Carter Doctrine laid the foundations for what would ultimately emerge as Centcom—the US military's Central Command. The American military is geographically divided into major regions, including the Southern Command, European Command, Pacific Command, and the Northern Command. In 1980,

---

48 Jim Harding, *After Iraq: War, Imperialism, and Democracy* (London: Merlin Press, 2004), 8-9.

49 Harding, *After Iraq*, 9-10.

50 Klare, *Blood and Oil*, 45.

51 Larry Everest, *Oil, Power, and Empire: Iraq and the U.S. Global Agenda* (Monroe, ME: Common Courage Press, 2003), 90.

in response to perceived Soviet expansion into the Middle East, Carter moved to project American force in the Gulf region. In particular, Carter established the Rapid Deployment Joint Task Force (RDJTF), a Tampa-based military command that would oversee US combat operations in the Gulf.

The Ronald Reagan administration, while highly critical of Carter's foreign policy approach, did endorse the basic premise of the Carter Doctrine.[52] Indeed, during the administration of Reagan the RDJTF would be elevated to a full-scale regional headquaters known as Centcom.[53] Established in January 1983, Centcom's 'area of responsibility' (AOR) covers twenty-five states in the Persian Gulf region, the Horn of Africa, the Caspian Sea basin, and Southwest Asia. The geographic region overseen by Centcom, not incidentally, contains the world's five leading producers of oil—Iran, Iraq, Kuwait, Saudi Arabia, and the United Arab Emirates—and many of its most important suppliers of natural gas.[54] Since its inception, Centcom forces have fought in four major conflicts: the Iran-Iraq War (1980-88); the Persian Gulf War (1991); the Afghanistan War (2001); and the Iraqi War (2003). Furthermore, almost all of the American soldiers who have died in combat since 1985 were serving under the authority of Centcom.[55]

Reagan likewise continued Nixon's policy of arming US allies in the region. Specifically, Reagan approved the largest US arms sale to Saudi Arabia ever: an US$8.5 billion package comprising 5 Airborne Warning and Control System (AWACS) surveillance aircraft, 7 KC-135 tanker aircraft, 660 Sidewinder air-to-air missiles, 22 ground radar installations, and a large array of air-defense and communications systems. Significantly, Congress balked at the sale of arms to Saudi Arabia. The House of Representatives voted to block the AWACS sale, and only through intensive appeals by Reagan did the Senate approve the deal. More so, Reagan's arming of Saudi Arabia carried with it another side. Reagan sought, and received, from the Saudi kingdom financial support for the CIA's clandestine campaigns to overthrow the Soviet-backed regimes in Afghanistan, Nicaragua, and elsewhere. Among the beneficiaries of this secret deal was Osama bin Laden, at the time a wealthy Saudi entrepreneur who helped recruit Islamic zealots to serve with the Afghan rebels. Klare contends that this business of war established a pattern of Saudi aid—'charitable contributions'— to militant Islamic groups in Afghanistan, and so laid the foundation for the rise of Al Qaeda and the Taliban in Afghanistan.[56]

Overt US foreign policy in the Middle East has always been relatively erratic, reflecting the vagaries of political fortunes and misfortunes. A constant thread, however, has been continued access to, and protection of, energy resources in the region. Nowhere is this better illustrated than in the on-again, off-again conflicts between Iran and Iraq. Throughout the twentieth century, most conflicts between

---

52 Klare, *Blood and Oil*, 46.
53 Klare, *Blood and Oil*, 46.
54 Klare, *Blood and Oil*, 2.
55 Klare, *Blood and Oil*, 2.
56 Klare, *Blood and Oil*, 47-48.

these two countries have resulted from discrepancies over land borders and fluvial boundaries. In 1979, for example, Saddam Hussein saw the fall of the Shah of Iran and the ensuring revolution as an opportunity. In the political turmoil, Saddam hoped to end the Islamic Republic's destabilizing agitation and perhaps even to topple the new regime before it had a chance to consolidate power. Moreover, Hussein saw this as an opportunity to capture Iran's southwest Khuzestan province, the heart of the country's oil industry. If successful, Iraqi oil production would increase from four to eleven million barrels a day, placing Baghdad in control of about 20 percent of the world's oil production and greatly increasing its global and regional leverage. Likewise, Iraq would gain possession of the Shatt al-Arab waterway. This channel is Iraq's only outlet to the sea; it also, though, provides access to Iran's oil refinery at Abadan. Following a 1937 treaty, Iraq assumed possession of the waterway. However, in 1969 the more powerful Iranian government took control of the Shatt al-Arab. Last, Iraq's decision to invade Iran was based on a re-configured geopolitical understanding. It was increasingly clear that the United States was becoming the most significant Western power in the region. Hussein, accordingly, sough through a mutual alliance against Iran to strengthen Iraq-American relations.[57]

In the aftermath of the Iranian coup, and the subsequent abduction of 52 Americans, the United States did reverse its position and support Iraq in its invasion. Throughout the ensuing eight-year war, the United States funneled massive amounts of military arms, intelligence, and aid to Iraq. Crucially, in 1982 the Reagan administration removed Iraq from the list of countries labeled as supporters of terrorism—despite knowledge by US officials that Saddam was employing chemical weapons against its Kurdish population and that Iraq was providing refuge to suspected Palestinian terrorists. Two years later the Reagan administration even restored full diplomatic relations between Iraq and the United States; Iran, however, was added to the list of terrorist states. As a result, Iran was denied access to American products and was severely limited in the procurement of arms delivered to it by other countries.[58]

In 1988 United Nations negotiators arranged for a cease-fire of the Iran-Iraq conflict. In retrospect, the war was indecisive. None of the major objectives specified by either Iran or Iraq were realized: neither regime was overthrown and the sovereignty of the Shatt al-Arab waterway remained in doubt. More significant is that the infrastructure of the two countries was severely damaged, particularly in the aftermath of the 'war of the cities' in which both Iraq and Iran launched missile attacks on heavily populated cities. Estimates place the number of military deaths for Iran and Iraq at 260,000 and 105,000 respectively; civilian deaths, however, were believed to be over one million.

Jim Harding is especially harsh in his summary of Western geopolitical intervention in the Middle East: "Through their support of coups and assassinations to protect Cold War and oil interests, the US and Britain contributed to the political culture

---

57   Simons, *From Sumer to Post-Saddam*; Everest, *Oil, Power, and Empire*, 95.
58   Simons, *From Sumer to Post-Saddam*, 322.

of violence within which Hussein rose to power."[59] Both Geoff Simons and Larry Everest concur, noting that the US support of dictators such as Saddam Hussein was influenced, partially, out of Cold War ideologies. This ideological tension, however, was far from abstract. The underlying objective has always been to retain Western access to, and control of, Middle Eastern energy resources.[60]

Apart from the Iran-Iraq conflict, Israel continued to occupy the attention of the Reagan administration. Since 1948 Israel has been viewed by many policy-makers as a reliable ally and a democratic bridgehead in an otherwise largely anti-American, undemocratic region. It has also been seen as a regional police officer and potential US surrogate that could be used to keep pro-Soviet radical states in line.[61] This is seen, for example, in the Lebanese civil war that began, after years of escalation between various religious, political, and clan-based factions, in 1975. Internal war erupted in Lebanon between Christians and Muslims, as well as between Israeli and Syrian troops. Israeli forces soon laid siege to the capital, Beirut. During the war Israeli forces entered the country, in 1978, and illegally occupied a strip of southern Lebanon. The Israelis viewed the existence of the PLO as a major threat to the legitimacy of its occupation of Palestinian land, and much of Israel's anti-PLO strategy in Lebanon involved trying to turn the Lebanese against the Palestinians.[62]

Foreign troops, including American, French, and Italian forces, were sent in to supervise a Palestinian Liberation Organization (PLO) evacuation. However, once this multinational contingent departed, Israeli troops returned. In 1982 Israel again invaded Lebanon ostensibly to destroy bases used by PLO guerrillas that had been attacking Israeli settlements. On 16 September 1982 Israeli forces began firing on thousands of unarmed children, women, and old met. The Sabra-Shatila Massacre resulted in between 2,000 and 3,000 Palestinian deaths and quickly transformed public perceptions about the war. Indeed, four hundred thousand Israelis marched in protest.[63]

Coalition forces were once again sent to the country in an attempt to stabilize the region. An ineffective United Nations tried to intervene but was unable to bring about peace. Many Lebanese, however, saw the arrival of thousands of American and other Western troops as a form of occupation. This multi-national force numbered 3,800 and was backed by a naval armada of American and French aircraft fighters, destroyers, and patrol boats.[64]

Starting in November 1982 Hezbollah, a loose federation of militant Shia groups that emerged in Lebanon in the early 1980s, carried out the first of 36 suicide terrorist attacks against American, French, and Israeli political and military targets in

---

59 Harding, *After Iraq*, 11.

60 Simons, *From Sumer to Post-Saddam*; Everest, *Oil, Power and Empire*.

61 Jordan, et al., *American National Security*, 401.

62 Phyllis Bennis, *Before & After: US Foreign Policy and the War on Terrorism* (New York: Olive Branch Press, 2003), 50.

63 Bennis, *Before & After*, 53.

64 Bennis, *Before & After*, 54.

Lebanon through 1986. On 18 April 1983, for example, the US embassy compound was destroyed by a car bomb, killing 63 people, of whom 17 were Americans, and wounding 100 more.[65] All together, these attacks killed 659 people.[66] As conditions in Lebanon deteriorated, Pentagon officials advised Reagan to withdraw. The president, however, feared a loss of political prestige in the event of a pull-out. Indeed, on 19 September 1983 Reagan ordered the shelling of Beirut from US naval ships. On 23 October 1983, however, an Islamic suicide bomber crashed a truck filled with explosives into a US Marines barrack at the Beirut airport. In the attack, 241 Americans and 58 French died. Four months later, after intense fighting, Reagan announced that the Marines in Beirut would withdraw.[67]

The demise of the Soviet Union presented American policy-makers with a problem vis-a-vis Middle East policy. Phyllis Bennis explains that without the Soviet Union, it would be more difficult for Washington to justify its aggressive assertion of regional power in the Middle East; however, the vital US interests—access to oil, defense of Israel, stability to encourage investment and market development—remained.[68] Such was the context of the 1991 Gulf War.

Throughout the 1980s the United States supported, albeit reluctantly, the Hussein-led Ba'ath Party. The rationale for this support was a combination of capitalist imperatives and Cold War practicality. However, following the collapse of the communist bloc, the Cold War justification dissolved. As a result, the US was in a position to re-evaluate its Middle Eastern ties. Saddam Hussein, for his part, faced economic uncertainty. Iraq's economy had been badly damaged during the inconclusive war with Iran. Any chance at rebuilding his country was predicated on oil revenues. However, Kuwait disregarded OPEC-established quotas, thereby depressing oil prices. Hussein claimed that this practice amounted to a form of economic warfare. Ironically, both Saudi Arabia and Kuwait had flooded the oil market during the Iran-Iraq war in an effort to damage Iran's economy. Now this same economic weapon, as charged by Sadam Hussein, was being leveled against Iraq. Evidence also suggests that the United States was complicit with Kuwait in the manipulation of oil prices.[69]

Saddam Hussein further justified his invasion of Kuwait on the grounds that Kuwaiti territory—acquired through the delineation of the region's modern political boundaries in the 1920s—rightly belonged to Iraq. Additionally, Saddam Hussein accused Kuwait of 'slant drilling' into Iraq's Rumeila oil field. And last, Hussein had conducted his war against Iran on the pretext of a Pan-Arab platform. Following the 1988 cease-fire Hussein assumed, or hoped, that neighboring Arab states, including

---

65   Bennis, *Before & After*, 55.

66   Robert A. Pape, *Dying to Win: The Strategic Logic of Suicide Terrorism* (New York: Random House, 2005), 129-131.

67   Bennis, *Before & After*, 55.

68   Bennis, *Before & After*, 59.

69   Simons, *From Sumer to Post-Saddam*, 343.

Kuwait and Saudi Arabia, would forgive Iraq of its debts. These requests, however, went unheeded.

Hussein also presumed that the United States would not interfere in his Middle East affairs. His assessment of the situation was based, in part, on numerous meetings with various US senators. During the early months of 1990, for example, Hussein met personally with Senators Robert Dole, Alan Simpson, Howard Metzenbaum, James McClure, and Frank Murkowski. Lastly, given the establishment of US-Iraqi business ventures, the previous American support of Iraq during the war against Iran, and the refusal of the George H.W. Bush administration to impose sanctions against Iraq (in light of human rights abuse charges), Hussein most likely concluded that Iraqi military intervention was acceptable with US geopolitical and economic interests. That said, it is also plausible that the Iraqi leader was in effect 'entrapped' by an ambiguous American foreign policy.[70]

In 1990 Iraqi military forces invaded and occupied Kuwait. In response, however, the United States did not turn a blind eye to Hussein's actions. Instead, the Bush Sr. administration banned all trade to Iraq and advanced UN Security Resolution 660. This latter resolution condemned the Iraqi invasion of Kuwait, demanded that Iraq withdraw immediately and unconditionally, and mandated that immediate negotiations be held between Iraq and Kuwait. At the same time, the Bush Sr. administration began to assemble a coalition in support of military intervention to liberate Kuwait.

Through a combination of punishments and rewards, American officials pieced together an international alliance of both Arab and Western governments. The US government, for example, forgave US$14 billion worth of Egyptian debts owed to the World Bank; promised military assistance and other forms of aid to Turkey and Syria—despite the fact that Syria was still recognized by US officials as a supporter of terrorism—and sent billions of dollars of aid to the failing Soviet Union.[71]

The Gulf War was short in duration, but long in consequences. In January 1991 US and British planes launched around-the-clock attacks on Baghdad; this air war lasted for six weeks. The subsequent ground invasion to 'liberate' Kuwait lasted only a few days. When the war was over, the occupation of Kuwait ended, Iraq lay in ruins, and the US had more power than ever in the region.[72] The only question that remained for American policy-makers was how to 'benefit' from America's emergence as the lone superpower and the preeminent power in the Middle East. According to Jordan and his colleagues, "The Persian Gulf War became a precipitating event for a new era, both for the region and the world as a whole. For the first time, the United States became massively involved militarily on the ground in the Middle East, thus becoming a significant part of the balance of power in the Persian Gulf."[73]

---

70 Lawrence Freedman and Effraim Karsh, *The Gulf Conflict 1990-1991: Diplomacy and War in the New World Order* (Princeton, N.J.: Princeton University Press, 1993).

71 Simons, *From Sumer to Post-Saddam*, 355-57.

72 Bennis, *Before & After*, 61-62.

73 Jordan et al., *American National Security*, 396.

The American victory over Iraq ushered in a 'New World Order' that was anything but clear. The sudden collapse of the Soviet Union resulted in a radical reorientation of Eastern Europe. Following the massive arms expenditure of the Reagan era, however, the US had moved from being a creditor state to being the largest debtor state in the world. As such, Bush Sr. could not offer the new emerging democracies in Eastern Europe anything like the Marshall Plan. Neither, for that matter, did Bush Sr. believe that the US should become the world's policeman.[74]

Pressure, however, continued to be directed toward Iraq. The Bush Sr. administration imposed an air and sea blockade, with the intended goal of preventing Saddam's regime from obtaining new arms or the technology to manufacture weapons of mass destruction. The Bush administration also established a permanent American military presence in Kuwait, declared a no-fly zone over southern Iraq, and 'pre-positioned' vast quantities of arms and ammunition at supply depots in Kuwait and Qatar.[75]

When William J. Clinton assumed the presidency in 1993 American foreign policy was at a cross-roads. Promising a bold new approach, Clinton said that his priorities would be to restore the American economy to good health—a prerequisite for foreign policy; to increase the importance attached to trade and open markets for American business; to demonstrate US leadership in the global economy; to help developing countries grow faster; and to promote democracy in Russia and elsewhere.[76] American policy in the Middle East remained focused primarily on Iraq and was largely consistent with the approach adopted by the Bush Sr. administration. Through the continued imposition of harsh sanctions and periodic bombings, the United States continued to apply pressure on the Hussein regime. Effectively, however, these actions only intensified the hold that Hussein and the Ba'ath Party had over Iraq. Concurrently, other European states, notably France and Russia, called for an easing of sanctions while continuing to enforce Iraq's disarmament. These demands had less to do with concern for the suffering people of Iraq than with the prospect of enlarging their own market opportunities in the country.

Clinton continued the policy of capitalist expansion through the promotion of democracy and overseas markets. However, unlike the 'containment' strategy of the Cold War, the Clinton administration promoted the concept of 'enlargement' as the prime goal of its foreign policy. Under this approach, the Clinton administration aimed to achieve rapid economic growth through overseas trade and investment—in effect, an unreserved embrace of predatory globalization, with its adoption of a neoliberal version of minimally regulated capitalism.[77]

On the surface, this approach is strikingly similar to the subsequent administration of George W. Bush. However, according to Neil Smith, "the Clinton administration's

---

74　Fraser Cameron, *US Foreign Policy after the Cold War: Global Hegemon or Reluctant Sheriff?* (New York: Routledge, 2002), 16-17.

75　Klare, *Blood and Oil*, 53.

76　Cameron, *US Foreign Policy*, 19.

77　Richard Falk, *The Great Terror War* (New York: Olive Branch Press, 2003), xiii.

power was thoroughly rooted in financial capital and, as such, it gave pure expression to a globalism innate to that sector. This was typified by a global neoliberalism that sought to establish, deregulate, and open up financial as well as commodity markets." The power of the Bush-Cheney presidency differs, however, in that "the list of companies benefitting from the multi-billion dollar publicly funded rebuilding of Iraq and the list of companies meeting with Cheney to design tax-funded corporate welfare for energy capitalists suggests that a class shift occurred with the accession of that administration." Smith argues that "the latter's social power pivots on the nexus between energy and the military and is rooted not in financial capital as such but in corporate capital devoted to the production of oil, energy equipment, armaments, aeronautics, military hardware and so forth."[78] The Bush administration, truly, was in the business of war. This becomes patently clear in the formulation of the Bush Doctrine and it is this shift that accounts for the emergence of a 'new' imperialism.

The Bush Doctrine and Permanent War

Speaking before the United States Military Academy, Bush announced in June 2002 that the United States was within its rights to use any and all means to ensure the security of its territorial integrity and of all its citizens. The 'Bush Doctrine' became official policy in September of that year when the White House released its 2002 *National Security Strategy*. The aggressiveness and unilateralism of the doctrine is made clear in the statement that the United States "will not hesitate to act alone, if necessary, to exercise [its] right of self-defense by acting preemptively ... against terrorists." Within the document, Bush wrote, "As a matter of common sense and self-defense, American will act against such emerging threats before they are fully formed. We cannot defend America and our friends by hoping for the best. So we must be prepared to defeat our enemies' plans, using the best intelligence and proceeding with deliberate action. History will judge harshly those who saw this coming danger but failed to act. In the new world we have entered, the only path to peace and security is the path of action."[79] These sentiments would be often repeated in the coming months. In late August, 2002, Vice President Cheney applied the new preemptive strike and unilateralist doctrine to Iraq, arguing, "What we must not do in the face of mortal threat is to give in to wishful thinking or willful blindness ... The risks of inaction are far greater than the risks of action."[80]

Although announced and codified in 2002, the Bush Doctrine was actually years in the making. In 1992 Paul Wolfowitz, then undersecretary of defense for policy, drafted a document for then Secretary of Defense Richard Cheney. This 46-page

---

78 Smith, *Endgame*, 20-21.

79 *National Security Strategy of the United States*, 2002, www.whitehouse.gov/nsc/nssall.html (March 31, 2004).

80 Quoted in Douglas Kellner, *From 9/11 to Terror War: The Dangers of the Bush Legacy* (Lanham, MD.: Rowman & Littlefield, 2003), 20.

draft document raised the possibility, among other things, of using American military power to preempt or punish the use of weapons of mass destruction *even in conflicts that otherwise do not directly engage US interests.* Moreover, the central strategy of the US Defense Department was to "establish and protect a new order" that accounts "sufficiently for the interests of the advanced industrial nations to discourage them from challenging" the leadership of the United States, while at the same time maintaining a military dominance capable of "deterring potential competitors from even *aspiring to a larger regional or global role.*" In other words, the document drafted by Wolfowitz advocated, first, an *unchallenged* American hegemony in the world and, second, the use of preemptive military force to maintain this position.[81] As summarized by Clark, this strategy for US global dominance required a hybrid economic/military/intelligence nexus in order to enforce American supremacy in the immediate post-Cold War period.[82]

The Wolfowitz report was leaked to the press and generated a fire-storm of criticism. Senator Robert Byrd (D-W.Va), for example, called the Pentagon's strategy "myopic, shallow and disappointing." Furthermore, according to Byrd, "The basic thrust of the document seems to be this: We love being the sole remaining superpower in the world and we want so much to remain that way that we are willing to put at risk the basic health of our economy and well-being of our people to do so." Likewise, Senator Joseph Biden (D-Del) ridiculed the document as "literally a Pax Americana." Biden explained that one cannot both be a lone superpower and still be able to maintain peace throughout the world."[83]

The Wolfowitz draft document was subsequently revised on paper; the final Defense Department report excised the 'offending' remarks about unchallenged US supremacy and global domination. These ideas, however, remained firm in the minds of key neoconservatives, including Wolfowitz and Cheney. This would become patently manifest in 1997 with the foundation of the Project for the New American Century (PNAC), a Washington D.C.-based think-tank. Launched by William Kristol, editor of the Robert Murdoch-owned neoconservative *Weekly Standard*, the PNAC has been influential in shaping the direction of US foreign policy. Its founding statement rebuked the presidency of William Clinton and called for a return to the elements of the earlier Ronald Reagan administration:

> We seem to have forgotten the essential elements of the Reagan administration's success: a military that is strong and ready to meet both present and future challenges; a foreign

---

81 Barton Gellman, "Keeping the U.S. First; Pentagon Would Preclude a Rival Superpower," *Washington Post*, March 11, 1992. www.yale.edu/strattech/92/dpg.html (August 25, 2005); emphasis added.

82 William R. Clark, *Petrodollar Warfare: Oil, Iraq and the Future of the Dollar* (Gabriola Island: New Society, 2005), 54.

83 Gellman, "Keeping the U.S."

policy that boldly and purposefully promotes American principles abroad; and national leadership that accepts the United States' global responsibilities.[84]

In 1998 the PNAC sent an open letter to Clinton. It bears mentioning that the policies of Clinton continued a trend of capitalist expansion through the promotion of democracy and overseas markets. The Clinton presidency, discursively, emphasized 'enlargement' as opposed to 'containment', as the primary goal of its foreign policy. Under this approach, the Clinton-era foreign policy was based on an attempt to achieve rapid economic growth through overseas trade and investment.[85] The Clinton administration, however, did not go far enough for the neoconservatives associated with the PNAC. For example, and according to the neoconservatives, Clinton failed to recognize that with the demise of the Soviet Union, the United States had been given a remarkable opportunity to assert American primacy. Moreover, these critics believed that Clinton ensnared the country in multilateral frameworks that did not serve broader international interests.[86] In reality, these interests were American-defined, American-motivated, and American-centered. Manifest destiny indeed figured prominently in the blueprints sketched out by the neoconservatives.

One of the first opportunities, according to those associated with the PNAC, was to promote regime change in the Middle East. In their open letter, the signatories encouraged Clinton to employ a preemptive strike against Iraq. The letter states:

> We are writing you because we are convinced that current American policy toward Iraq is not succeeding, and that we may soon face a threat in the Middle East more serious than any we have known since the end of the Cold War ... You have an opportunity to chart a clear and determined course for meeting this threat. We urge you to seize that opportunity, and to enunciate a new strategy that would secure the interests of the U.S. and our friends and allies around the world. That strategy should aim ... at *the removal of Saddam Hussein's regime from power*.[87]

In a call for unilateral military action, the letter continues: "We can no longer depend on our partners in the Gulf War coalition to continue to uphold the sanctions or to punish Saddam when he blocks or evades UN inspections." Furthermore, "if Saddam does acquire the capability to deliver weapons of mass destruction, as he is almost certain to do if we continue along the present course, the safety of American troops in the region, of our friends and allies like Israel and the moderate Arab states, and a significant portion of the world's supply of oil will all be put at hazard."[88] The letter was signed by eighteen members of the PNAC, including Donald Rumsfeld, Paul Wolfowitz, Peter Rodman, William Schneider, Richard Perle, Richard Armitage,

---

84 Alex Callinicos, *The New Mandarins of American Power: The Bush Administration's Plans for the World* (Cambridge, UK: Polity Press, 2003), 50.

85 Falk, *Great Terror War*.

86 Daalder and Lindsay, *America Unbound*, 12.

87 Project for the New American Century, "Open Letter to President Clinton," www.newamericancentury.org/iraqclintonletter.htm (April 27, 2004).

88 Project for the New American Century, "Open Letter".

Paula Dobriansky, John Bolton, Zalmay Khalilzad, and Elliott Abrams. Most of these individuals would assume prominent positions in the administration of George W. Bush.

In September 2000, in the months leading up to the presidential election between Bush and Al Gore, the PNAC released a major policy study entitled *Rebuilding America's Defenses: Strategy, Forces and Resources for a New Century*.[89] The report is a massive blueprint for an American empire. The PNAC explicitly built upon the defense strategy outlined by the Cheney Defense Department—and spearheaded by Wolfowitz—during the final days of the Bush Sr. administration. As expressed in the 2000 report, "The Defense Policy Guidance (DPG) drafted in the early months of 1992 provided a blueprint for maintaining US preeminence, precluding the rise of a great power rival, and shaping the international security order in line with American principles and interests." That report, having come under harsh international criticism for its idea of *Pax Americana* was "buried" by the Clinton administration. The contributors to the 2000 report, however, found that "the basic tenets of the DPG ... remain sound."[90]

The PNAC 2000 report laid out their proposed vision for America in the 21st century: "At present the United States faces no global rival. America's grand strategy should aim to preserve and extend this advantageous position as far in to the future as possible."[91] Further, the function of the US military "is to secure and expand the 'zones of democratic peace;' to deter the rise of a new great-power competitor; defend key regions of Europe, East Asia, and the Middle East; and to preserve American preeminence through the coming transformation of war made possible by new technologies."[92] The report likewise acknowledged that "While the unresolved conflict with Iraq provides the immediate justification, the need for a substantial American force presence in the Gulf transcends the issue of the regime of Saddam Hussein."[93]

As William Engdahl notes, there was no talk of Iraqi's weapons of mass destruction, or of any ties between the regime of Saddam Hussein and terrorists. Indeed, Engdahl emphasizes that "Months before the world ... witnessed the attacks on the World Trade Center and the Pentagon, or had even heard of Osama bin Laden, Cheney's PNAC had targeted Saddam Hussein's Iraq for special treatment, stating bluntly that US policy should be to take direct military control of the Arabian Gulf."[94]

---

89 Project for the New American Century, *Rebuilding America's Defenses: Strategy, Forces and Resources for a New Century*, www.newamericancentury.org/RebuildingAmericasDefenses.pdf (April 27, 2004).

90 Project for the New American Century, *Rebuilding America's Defenses*, ii.

91 Project for the New American Century, *Rebuilding America's Defenses*, i.

92 Project for the New American Century, *Rebuilding America's Defenses*, 2-3.

93 Project for the New American Century, *Rebuilding America's Defenses*, 14.

94 William Engdahl, *A Century of War: Anglo-American Oil Politics and the New World Order*, revised edition. (London: Pluto Press, 2004), 252.

Foreshadowing Bush's 'Axis of Evil', the report singled out three states—North Korea, Iran, and Iraq—as posing threats to both American leadership and America itself. That the 2000 PNAC was prescient should come as no surprise, given the composition of the PNAC. Indeed, all of these early reports and communiques should be seen as providing the foundation for the actions of the Bush administration.[95]

Operationally, the Bush Doctrine rests on two beliefs. The first argues that the best, if not only, way to ensure America's security is to shed the Wilsonian ideals of multinational alliances and international institutions; the second is that America should use its strength to change the status quo in the world.[96] In short, America should use its unrivaled position to remake the world in its own image. The Bush administration proclaimed that it would no longer be bound to international law or institutions. In reference to the Bush Doctrine, Larry Everest concludes that "No other empire in history had ever issued such an arrogant, blatant, and chilling declaration of global hegemony."[97] Although critics of the foreign policies of Monroe, Polk, Roosevelt, and Eisenhower may disagree, the Bush Doctrine does chart an exceptionally unilateral and militaristic course for American foreign policy. As summed by Daalder and Lindsay, Bush "relied on the unilateral exercise of American power rather than on international law and institutions to get his way. He championed a proactive doctrine of preemption and de-emphasized the reactive strategies of deterrence and containment. He promoted forceful interdiction, preemptive strikes, and missile defenses as means to counter the proliferation of weapons of mass destruction, and he downplayed America's traditional support for treaty-based non-proliferation regimes. He preferred regime change to direct negotiations with countries and leaders that he loathed. He depended on ad hoc coalitions of the willing to gain support abroad and ignored permanent alliances. He retreated from America's decades-long policy of backing European integration and instead exploited Europe's internal divisions."[98]

Moreover, the Bush Doctrine does not, in fact, stop at 'preemption.' As William Hartung explains, the United States has engaged in military 'first strikes' in the past, including, for example, the 1989 invasion of Panama. What the Bush Doctrine does, however, is to elevate preemption from an occasional tactic to a guiding principle of US foreign policy. Hartung concludes that "The notion of attacking another nation only in legitimate self-defense, a well-established principle of international law that has been violated far too often as it is, would be seriously undermined if the preemption doctrine were to become an enduring component of US foreign policy."[99]

---

95 Project for the New American Century, *Rebuilding America's Defenses*, 75.
96 Daalder and Lindsay, *America Unbound*, 13.
97 Everest, *Oil, Power & Empire*, 27.
98 Daalder and Lindsay, *America Unbound*, 2.
99 William D. Hartung, "Military," In *Power Trip: U.S. Unilateralism and Global Strategy after September 11*, edited by John Feffer (New York: Seven Stories Press, 2003), 69-74; at 66-67.

Part of American foreign policy, moreover, is a move toward an Orwellian state of permanent war. Boal and his co-authors contend that "Regardless of the tactical details of various invasions, occupations, and retreats, the US empire has followed a long and consistent strategic path—centered on and driven by military engagement—to force regional penetration and exploit the existing or resulting 'weak states'. And throughout this history, certainly no less so over the past several decades, what passes for 'peace' has prefigured—has been *structured* to prefigure—an endless series of wars."[100]

Ironically, a move toward permanent war has been—and continues to be—promoted as a path toward peace and stability. Faced with a new borderless, transnational enemy, the preemptive and unilateralism evinced in the Bush Doctrine is frequently justified as a sound response to a 'new world order'. The dangers of 'rogue' states and 'failed' states, along with transnational drug cartels, arms dealers, and terrorist networks are constantly evoked to instill fear. The emergence of these transnational dangers, however, is rarely spelled out by public officials. Instead, Bush and his administration declared that freedom and fear are at war, and that the advance of human freedom is at stake.

Freedom, however, is not in jeopardy—unless one considers the increasing state surveillance and illegal wire-tapping activities to ensure the perpetuation of the status-quo. Rather, the Bush Doctrine, and the subsequent war in Iraq, are the culmination of a series of neoliberal policies and transnational practices pursued in the spirit of Pax Americana. The shock of 11 September 2001 and the aftermath of mass patriotism, infringements of civil liberties and homeland security simply forged the conditions to act on the unilateralist militarist perspective.[101] According to Henry Giroux, "Embracing a policy molded largely by fear and bristling with partisan, right-wing ideological interests, the Bush administration took advantage of the tragedy of 9/11 by adopting and justifying a domestic and foreign policy that blatantly privileged security over freedom, the rule of the market over social needs, and militarization over human rights and social justice."[102] As Boal and his associates explain, the "chaos of ongoing war" provides the justification for the "implantation of external economic forces: the pumping up of an addictive regional arms trade, and the military and its contractors acting as advance men for 'development' capital (e.g., pipelines, terminals, transport and communication systems), which gains a stranglehold on the weak states that emerge from the melee. In the longer term, warfare's devastation of infrastructure requires an infusion of handouts and investment with various neo-liberal conditions attached. In all of these ways, war's service to capital is to set the stage for the trinity of crude accumulation:

---

100 Boal et al., *Afflicted Powers*, 93.

101 Harding, *After Iraq*, 36.

102 Henry Giroux, *The Terror of Neoliberalism* (Boulder, CO.: Paradigm Publishers, 2004), 2.

the enclosure and looting of resources; the creation of a cheap and deracinated labor force; and the establishment of captive markets."[103]

Off to War

Presidents do not directly ask Americans to sacrifice their lives for corporate greed. Nor are Americans asked to die for the grotesque profits garnered by politicians. As CEO of the world's largest oil-and-gas-services company, for example, Vice-President Cheney earned $44 million in salary from Halliburton—a corporation that on his own admission saw war as a 'growth opportunity.'[104] Rather, Americans are called to pay the ultimate price in defense of ambiguous—though no less powerful—phrases such as freedom, democracy, and liberty. Nicholas Lemann details, through an examination of key speeches, how the Bush administration gradually enlarged the immediacy of the 9/11 attacks to conform with the overriding goals of the PNAC blueprint for US hegemony.[105] Immediately following the attacks of 11 September 2001, for example, Bush declared "The search is underway for those who are behind these evil acts....We will make no distinction between the terrorists who committed these acts and those who harbor them." He concluded that "America and our friends and allies join with all those who want peace and security in the world, and we stand together to win the war against terrorism."[106] Nine days later, in an address delivered before the US Congress, Bush identified the perpetrators of the attacks as Al Qaeda. The American response to the attacks, according to Bush was to defeat Al Qaeda and those governments found to support and harbor members of Al Qaeda. The Taliban regime of Afghanistan was explicitly singled out, although Bush also noted that Al Qaeda operatives were found in more than 60 countries. Bush stated that "Our war on terror begins with Al Qaeda, but it does not end there. It will not end until every terrorist group of global reach has been found, stopped and defeated." He continued that "we will pursue nations that provide aid or safe haven to terrorism. Every nation, in every region, now has a decision to make. Either you are with us, or you are with the terrorists. From this day forward, any nation that continues to harbor or support terrorism will be regarded by the United States as a hostile regime."[107] Lemann maintains that Bush was broadening the United States' understanding of being at war, extending it from international terrorist organizations to governments that were

---

103 Boal et al., *Afflicted Powers*, 99-100.

104 Boal et al., *Afflicted Powers*, 41.

105 Nicholas Lemann, "The War on What?" *The New Yorker*, September 16, 2002, www.newyorker.com/printables/fact/020916fa_fact (August 30, 2005).

106 Office of the Press Secretary, "Statement by the President in his Address to the Nation," September 11, 2001, www.whitehouse.gov/releases/2001/09/print/20010911-16.html (July 27, 2004).

107 Office of the Press Secretary, "Address to a Joint Session of Congress and the American People," September 20, 2001, www.whitehouse.gov/news/releases/2001/09/print/20010920-8.html (July 27, 2004).

not necessarily connected to Al Qaeda or even involved in the 11 September attacks. The Bush administration, in the words of William Clark, "appears to have made a coordinated attempt to create massive societal fear."[108]

Three primary and interlocking fear-based explanations were used to justify the invasion and occupation of Iraq. The Bush administration's initial justification for war was predicated on the assertion that Saddam Hussein and his Ba'athist Regime possessed weapons of mass destruction (WMD). Second, justification hinged on the assertion that even if Saddam did not use these weapons against the United States or its friends and allies, then Saddam would sell them to terrorists networks such as Al Qaeda. Third, the Bush administration forwarded the idea that the Iraqi government had definitive ties to Al Qaeda and, by extension, to the attacks of 11 September. Consequently, the Bush administration justified military intervention on the grounds of the November 2002 UN Security Resolution 1441, which required Hussein to comply with UN disarmament obligations or face serious consequences. Only later, when confronted with luke-warm support for the war effort within the global community, did calls for the liberation of the Iraqi people arise.

On 26 August 2002 Cheney appeared at the 103rd National Convention of the Veterans of Foreign Wars (VFW). Cheney stated that the number one responsibility of the Bush administration was "to protect the American people against future attack, and to win the war [on terrorism] that began [on] September 11." He then detailed, in sweeping strokes, the on-going campaign against Al Qaeda and the Taliban regime. Next, however, in reference to earlier statements by Bush, Cheney indicated that America could only keep the peace by "redefining war" on its own terms. For Cheney, this meant that "any enemy conspiring to harm America or our friends must face a swift, a certain and a devastating response." Cheney explained that "the challenges to our country involve more than just tracking down a single person or one small group." In other words, America's response was no longer concerned with bringing to justice those immediately responsible for the 11 September attacks. Cheney said that "Nine-eleven and its aftermath awakened this nation to danger, to the true ambitions of the global terror network, and to the reality that weapons of mass destruction are being sought by determined enemies who would not hesitate to use them against us." Similar to Bush, Cheney was expanding the response to September 11 through the construction of a 'global terror network.' Discursively, this was portrayed as a massive conspiracy of inter-linked groups and governments, all of whom, in some fashion or another, contributed to the attacks.[109]

Having established the basis of a global network, Cheney abruptly introduced the 'case' of Saddam Hussein. Cheney explained that "The case of Saddam Hussein, a sworn enemy of our country, requires a candid appraisal of the facts." The regime of Saddam, according to Cheney, had been systematically enhancing its capabilities

---

108 Clark, *Petrodollar*, 107.

109 Office of the Press Secretary, "Vice President Speaks at VFW 103rd National Convention," August 26, 2002, www.whitehouse.gov/news/releases/2002/08/print/20020826. html (August 29, 2005).

of weapons of mass destruction, including nuclear weapons. Striking fear in his audience, Cheney warned: "Should all his [Saddam's] ambitions be realized, the implications would be enormous for the Middle East, for the United States, and for the peace of the world. The whole range of weapons of mass destruction then would rest in the hands of a dictator who has already shown his willingness to use such weapons, and has done so, both in his war with Iran and against his own people."[110] Cheney did not, however, explain that the United States knew of the chemical attacks during the Iraq-Iran War and, indeed, provided intelligence that Saddam utilized in the attacks. What Cheney did explain, though, was that "Armed with an arsenal of these weapons, and seated atop ten percent of the world's oil reserves, Saddam Hussein could then be expected to seek domination of the entire Middle East, take control of a great portion of the world's energy supplies, directly threaten America's fiends throughout the region, and subject the United States or any other nation to nuclear blackmail." Cheney declared simply: "there is no doubt that Saddam Hussein now has weapons of mass destruction. There is no doubt he is amassing them to use against our friends, against our allies, and against us. And there is no doubt that his aggressive regional ambitions will lead him into future confrontations with his neighbors—confrontations that will involve both the weapons he has today, and the ones he will continue to develop with his oil wealth."[111]

One month later, Secretary of Defense Donald Rumsfeld likewise informed the American public that "Some have argued that the nuclear threat from Iraq is not imminent—that Saddam is at least 5-7 years away from having nuclear weapons. I would not be so certain. And we should be just as concerned about the immediate threat from biological weapons. Iraq has these weapons." The following day Rumsfeld elaborated that "No terrorist state poses a greater or more immediate threat to the security of our people and the stability of the world than the regime of Saddam Hussein in Iraq."[112]

President Bush was equally vociferous in his condemnation of Saddam Hussein. Bush made repeated claims in the months leading up to war that the Iraqi dictator posed an imminent and grave threat to American security. In September 2002, for example, Bush stated that "This man [Saddam Hussein] poses a much graver threat than anybody could have possibly imagined." In October, he reaffirmed that "There's a grave threat in Iraq" and that "The Iraqi regime is a threat of unique urgency." Bush explained in more detail that "There are many dangers in the world, the threat from Iraq stands alone because it gathers the most serious dangers of our age in one place. Iraq could decide on any given day to provide a biological or chemical weapon to a terrorist group or individual terrorists." He concluded, in subsequent speeches, that

---

110 Office of the Press Secretary, "Vice President Speaks."

111 Office of the Press Secretary, "Vice President Speaks." Emphasis added.

112 "In Their Own Words: Iraq's 'Imminent' Threat. Center for American Progress. www. americanprogress.org/site/pp.aspx?c=biJRJ8OVF&b-24970 (August 29, 2005).

"There is a real threat, in my judgement, a real and dangerous threat to America ... in the form of Saddam Hussein" and that "Saddam Hussein is a threat to America."[113]

In the build-up to war, Bush and other members of his administration justified war through explicit links between Saddam's regime and Al Qaeda. During an eight-day period in 2002 (28 October to 4 November), for example, Bush claimed this association in eleven separate speeches.[114] On the eve of war, Bush declared in a televised speech on CNN that "Intelligence gathered by [the United States] and other governments leaves no doubt that the Iraq regime continues to possess and conceal some of the most lethal weapons ever devised."[115]

The Bush administration consciously and deliberately used the threat of terror as a justification for its long-standing goal of regime change in Iraq. The attacks of September 11 simply provided a 'just' cause. In the aftershock of 9/11, National Security Advisor Condoleezza Rice encouraged members of the National Security Council 'to capitalize on these opportunities to fundamentally change American doctrine' and Rumsfeld called for an immediate attack on Iraq."[116] As Jim Harding concludes, "It is now irrefutable that 9/11 was an excuse for the Bush administration to launch its Pax Americana campaign.

Resolution 1441 explicitly stated that only the Security Council could decide when and what steps may be taken for the implementation of the resolution. The Bush administration was not legally justified to act unilaterally in its quest for war. As Peter Singer summarizes, under Resolution 1441 the Security Council had retained the right to make that decision itself and to decide on the nature of the consequences that would follow if Iraq was found to be in violation of its obligations. This is the reason why the Bush administration tried so hard to obtain a second resolution declaring that Iraq had not disarmed and authorizing the use of force against Iraq. This also accounts for the reluctance of some governments to support the Bush administration's Coalition of the Willing.[117] The Bush administration, nevertheless, declared emphatically that it was justified to act preemptively.

But there never was an Iraqi threat to the security of the United States. In the weeks following the invasion and occupation of Iraq, members of the Bush administration attempted to cover their tracks. On 14 May 2003, for example, speaking before the US Senate's appropriations subcommittee, Rumsfeld declared that "I don't believe anyone that I know in the administration ever said that Iraq had nuclear weapons."[118] More egregious was the turn-around of Cheney. Just days before the invasion of Iraq

---

113 "In Their Own Words."

114 Peter Singer, *The President of Good and Evil: The Ethics of George W. Bush* (New York: Dutton, 2004), 257.

115 "Bush: 'Leave Iraq within 48 hours,'" transcript of Bush's televised speech to the nation, *CNN*, March 17, 2003, www.cnn.com/2003/WORLD/meast/03/17/sprj.irq.bush. transcript/ (August 29, 2005).

116 Harding, *After Iraq*, 145.

117 Singer, *President of Good and Evil*, 159.

118 Timothy Noah, "Whopper of the Week: Donald Rumsfeld, Meet Dick Cheney," *slate. msn.com*, May 23, 2003. http://slate.msn.com/id/2083532 (August 29, 2005).

was to begin, Cheney appeared on NBC's *Meet the Press* and said that "we believe he has, in fact, reconstituted nuclear weapons."[119] However, just seven months later on 14 September 2003, again on *Meet the Press*, Cheney agreed that he 'mis-spoke', and that the Bush administration "never had any evidence that he had acquired a nuclear weapon." Moreover, Cheney explained that Saddam *never* had any weapons of mass destruction, only that the Bush administration believed that Saddam was *attempting* to acquire the capabilities for a WMD program.[120] This is not simply a matter of mis-speaking. Rather, the repeated references by Bush, Cheney, and Rumsfeld contributed, on the one hand, to a general climate of fear and paranoia in the United States. On the other hand, these statements enunciated by the Bush administration were deliberately used to justify support for the US-led invasion and occupation of Iraq. The 'imminent' threat posed by the Saddam Hussein regime was magnified by its alleged connection to Al Qaeda and the attacks of 11 September 2001. On the 14 September 2003 episode of *Meet the Press*, for example, moderator Tim Russert asked of the Vice President: "The *Washington Post* asked the American people about Saddam Hussein, and this is what they said: 69 percent said he was involved in the September 11 attacks. Are you surprised by that?" Cheney responded, "No. I think it's not surprising that people make that connection." Russert asked in a follow-up, "But is there a connection?" The response was simple: "We don't know."[121]

There was no connection. Neither the CIA, the FBI, nor the 9/11 Commission could find any links between Saddam Hussein and Al Qaeda.[122] Indeed, it is highly unlikely that a secular dictator such as Saddam Hussein would provide any weapons to the Islamic fundamentalist. On the one hand, during the first Gulf War bin Laden volunteered his forces to help remove Saddam; on the other hand, Saddam embarked on war against Iran in an attempt to ameliorate the threat of Islamic fundamentalism. As Clark identifies, to suggest a relationship between Saddam Hussein, a secular despot who oppressed the religious Shi'ite majority in Iraq, and Al Qaeda terrorists who endorse a fanatical form of fundamentalist Islamic Whabbism defies both logic and that region's history.[123] And lastly, despite over 400 unfettered prewar UN inspections, no substantive evidence was ever reported to the Security Council that Iraq was reconstituting its WMD program.[124] Clarke concludes, "Even if Iraq still had WMD stockpiles, possession of weapons of mass destruction is not in and of

---

119 Wolf Blitzer, "Did the Bush Administration Exaggerate the Threat from Iraq?," *CNN. com*, July 8, 2003, www.cnn.com/2003/ALLPOLITICS/07/08/wbr.iraq.claims/ (August 29, 2005).

120 Vice President Cheney, transcript, Tim Russert, moderator, September 14, 2003, www.msnbc.msn.com/id/2080244/ (August 29, 2005).

121 Vice President Cheney, transcript.

122 Clark, *Petrodollar Warfare*, 98.

123 Clark, *Petrodollar Warfare*, 3.

124 Clark, *Petrodollar Warfare*, 3.

itself a threat to the United States. Over two dozen nations possess WMD, according to unclassified CIA testimony to Congress."[125]

And yet, even after the war 'ended', in his speech aboard the USS Abraham Lincoln, Bush continued to connect the war in Iraq with the September 11 attacks and Al Qaeda. Bush explained that "The battle of Iraq is one victory in a war on terror that began on September the 11, 2001—and still goes on. That terrible morning, 19 evil men—the shock troops of a hateful ideology—gave America and the civilized world a glimpse of their ambition." Bush reminded his audience that "We have not forgotten the victims of September the 11th—the last phone calls, the cold murder of children, the searches in the rubble. With those attacks, the terrorists and their supporters declared war on the United States. And war is what they got." In his next sentence, Bush reaffirmed the connection between the terrorist attacks and Iraq: "Our war against terror is proceeding according to the principles that I have made clear to all: Any person involved in committing or planning terrorist attacks against the American people becomes an enemy of this country, and a target of American justice." He stated that "The liberation of Iraq is a crucial advance in the campaign against terror. We've removed an ally of Al Qaeda, and cut off a source of terrorist funding." Emphatically, he concluded that "this much is certain: No terrorist network will gain weapons of mass destruction from the Iraqi regime, because the regime is no more."[126]

So what accounts for the death and destruction? George Leaman writes that it is abundantly clear that the Bush administration misrepresented its case for war with Iraq in an effort to win support from the American public, the US Congress, and the U.N. Security Council.[127] Clark, likewise, asserts that "it is irrefutable that this conflict's stated objectives were at best deceptive and at worst outright fraudulent." The Bush administration required an external threat in order to use overt military force to pursue ulterior objectives. The objectives, of course, are related to the perpetuation of an American empire. More concretely, these objectives facilitated the business of war.

The Business of War

Members of the Bush administration defined the war on their terms: preventing Saddam from developing a program of WMD; to prevent WMD from falling into the hands of terrorists; to rid the world of a dictator; and to liberate the Iraqi people and to promote democracy. Critics of the war have been no less prolific in their

---

125 Clarke, *Against All Enemies: Inside America's War on Terror* (New York: Free Press, 2004), 267.

126 Office of the Press Secretary, "President Bush Announces Major Combat Operations in Iraq Have Ended," 1 May 2003, www.whitehouse.gov/news/releases/2003/05 (3 August 2006).

127 George Leaman, "Iraq, American Empire, and the War on Terrorism," *Metaphilosophy* 35 (2004): 234-248; at 234.

identification of causes. The war, conversely, has been framed "as an exemplary instance of gunboat diplomacy in the interests of 'free trade'; as a consequence of the seizure of power by the Project for the New American Century; as a demonstration of the price to be paid by any state opposing the vision of the world order laid out in the National Security Strategy document of September 2002; as a road test for Rumsfeld's new model of the military; to permit the withdrawal of US troops from Saudi Arabia; to complete George H.W. Bush's unfinished business of the first Gulf War; as a spectacular response to September 11; even as a reaction to the lack of targets in Afghanistan."[128] To these we can add the following reasons: "To clean up the mess left by the first Bush administration when it let Saddam Hussein consolidate power and slaughter opponents after the first US-Iraq war; to improve Israel's strategic position by eliminating a large, hostile military; to create an Arab democracy that could serve as a model to other friendly Arab states now threatened with internal dissent, notably Egypt and Saudi Arabia; and to create another friendly source of oil for the US market and reduce dependency upon oil from Saudi Arabia, which might suffer overthrow someday."[129]

These rationales—which do not include any reference to the 'liberation' of the Iraqi people—are *apparently* united by a single underlying concern over the stability of energy resources in the Middle East. It is argued, consequently, that the war was a tradeoff: Blood for Oil.

Was the war conducted simply under the auspices of oil? The short answer is yes and no. For the modern state, oil is essential—and in high demand. In 2004 global oil consumption was estimated at 82 million barrels a day (mb/d). In 20 years, however, consumption is projected to increase to 120 mb/d. The United States, though, remains the world's largest energy consumer. In 2004, for example, with a consumption of 20 mb/d, the United States consumed nearly one out of every four barrels produced. This constitutes a significant case of hyper-consumption, in that the United States contains only five percent of the world's population.[130] It is not simply that the United States requires massive imports of oil. Other countries are also projected to require substantially more imports to maintain their economies. It is anticipated, for example, that by 2020, Asian economies led by China will consume 25 percent of the world's energy.[131] From the perspective of certain officials in the Bush administration, any increased demand elsewhere constitutes a threat not only to American supplies but, by extension, to American dominance in the global economy. Oil becomes a zero-sum game. You either have it, or you do not have it.

The oil explanation is based on three arguments: scarcity, dependency, and petrodollars. Combined these elements create a 'new political cartography of oil.'[132] First, petroleum is a scarce, finite resource and these levels of consumption

---

128 Boal et al., *Afflicated Powers*, 54.
129 Richard A. Clarke, *Against All Enemies*, 265.
130 Clark, *Petrodollar Warfare*, 76.
131 Clark, *Petrodollar Warfare*, 47.
132 Boal et al., *Afflicted Powers*, 57.

cannot continue indefinitely. The United States, as a case in point, reached peak oil production in 1970. By 2020 the United States will need to import 65 percent of projected oil demand. And recurrent fears of oil scarcity—and its twin fear, access—have influenced US foreign policy for the last six decades. Since the Second World War petroleum has been deemed vital to the security and prosperity of the United States. In 1943, for example, Roosevelt declared that "the defense of Saudi Arabia is vital to the defense of the United States." Saudi Arabia, at the time, was not under attack by any of powers, such as Germany or Japan; moreover, the United States had no prior ties with Saudi Arabia. What did exist, however, was a fledgling presence in the Middle East of American oil companies. Beginning in 1933 when the Standard Oil Company of California obtained a concession over vast tracts of land in the eastern province of Saudi Arabia, American geologists began to look to the Persian Gulf as a bonanza of future oil fields.[133] In subsequent years, various presidential administrations collaborated with giant America oil companies, facilitating the latter to obtain concessions in the region and providing them with diplomatic and military support. Moreover, during the 1950s petroleum took on more strategic significance as the United States and other major powers came to rely on cheap and abundant oil to fuel their booming economies and, in the case of the United States, its military.[134]

Fears of scarcity became increasingly more pronounced during the 1970s and 1980s. Throughout the latter decades of the twentieth-century, America's petroleum supply was increasingly imported, and especially imported from the Gulf. The share of America's petroleum supply that is accounted for by imports reached 30 percent in 1973 and 40 percent in 1976; in 1998 American dependence on imported petroleum crossed the 50 percent mark.[135] Scarcity, however, is difficult to assess for oil reserves. Oil is an item "of market currency, and therefore subject to constantly shifting expectations and perceptions, speculation and gambling". The price hikes throughout the oil crises of 1973-74 and 1979-80, for example, had nothing to do with actual oil scarcity.[136] Indeed, an historical review indicates that oil policies have been designed to maintain a system of organized scarcity capable of keeping oil prices low enough for capitalist growth and high enough for corporate profitability.[137]

The specter of scarcity leads to problems of access. Given the uncertainty of the Gulf in its ability to meet future oil demand, the Cheney report encouraged the diversification of other supplies. Russia and the major Caspian Sea producers—Azerbaijan, Kazakhstan, Turkmenistan, and Uzbekistan were notably singled out as promising areas. After the breakup of the Soviet Union in 1991, five independent states suddenly bordered the Caspian Sea: Russia, Iran, Azerbaijan, Kazakhstan, and Turkmenistan. Although it is unclear as to the size of oil and gas fields in the region, some estimates suggest that Azerbaijan, Kazakhstan, and Turkmenistan

133 Klare, *Blood and Oil*, 30-2.
134 Klare, *Blood and Oil*, 37.
135 Klare, *Blood and Oil*, 13.
136 Boal et al., *Afflicted Powers*, 61, 64.
137 Boal et al., *Afflicted Powers*, 60-61.

alone may have proven reserves of just under eight billion barrels, but that possible reserves may reach over 200 billion barrels.[138] During the 1990s American-based multinational oil companies, followed by the US military, moved in to secure oil and gas resources of the region.[139] This area is geographically sensitive in that none of these states has access to the world's oceans, and thus the need to construct and maintain pipelines. Oil and gas must be transported to market via exposed pipelines and hence require military protection. It is not surprising, therefore, to discover that many US military operations in the 'War on Terror' have been conducted in critical sites associated with the production and transfer of oil. In February 2002, for example, approximately 150 Special Forces and ten combat helicopters were dispatched to the Republic of Georgia in the Caucasus. The cover story for the operation was that they were preparing Georgian forces to fight Chechen rebels with alleged Al Qaeda connections. In actuality, these forces were deployed to protect the Baku-Ceyhan pipeline. Construction for this project began in 2002; once completed, the 1,091-mile-long-conduit would stretch from Baku, through Azerbaijan, through Georgia to its Black Sea port of Batumi, and then across Turkish Kurdistan, to Turkey's deep-water port of Ceyhan.[140]

Apart from retaining access to oil fields, American officials are also concerned with preventing access to other states. China, for example, has attempted to negotiate a possible pipeline from Kazakhstan to Shanghai via Xinjiang Province; additionally, China is also attempting to obtain oil from Russia via a pipeline that would stretch from Angarsk in Siberia to the Daqing oil field in Manchuria.[141] And not to be ignored, during the 1990s various 'foreign suitors' were quietly pursuing their own access to the Iraqi oil fields. Indeed, Cheney and the National Energy Group reviewed several documents in the spring of 2001, including 'Foreign Suitors for Iraqi Oilfield Contracts'. They identified that by 1997 many states, including China, France, Russia, and even Japan, had approached Saddam Hussein to secure oil exploration contracts. French and Russian companies, among others, had established 'production-sharing contracts' with Iraq. An international consensus— and not one shared by the United States— had developed by the mid- to late-1990s that Iraq's WMD program had been effectively dismantled by economic sanctions. It was assumed by these governments that once sanctions were lifted, they would profit through the awarding of oil-lease contracts by the Iraqi government. The United States would be left out.[142] In part, this explains why the United States under the presidential administration of Clinton, pulled UN weapons inspectors out of Iraq in 1998 and launched a series of bombing campaigns.

Fears over continued access to (cheap) oil are accompanied by the dangers of foreign dependence on oil. This constitutes the second coordinate in the new

---

138 Johnson, *Sorrows of Empire*, 169.
139 Johnson, *Sorrows of Empire*, 169.
140 Johnson, *Sorrows of Empire*, 174-5.
141 Johnson, *Sorrows of Empire*, 170.
142 Clark, *Petrodollar Warfare*, 62.

political cartography of oil. Klare writes that "abundant petroleum has helped the US economy and the US military dominate the world, and to propel further growth [the US] will have to consume more and more of it, yet the United States is producing less oil, and thus will have to import ever increasing quantities from abroad. The United Kingdom, likewise, has no significant oil reserves other than the North Sea. This field, furthermore, reached peak oil production in 1999. It is speculated that one reason why British Prime Minister Tony Blair so ardently supported the invasion of Iraq was because Britain had passed its own peak of oil production.

And herein lies a dilemma: "Oil makes the United States strong; dependency makes it weak."[143] Specifically, Klare suggests that dependency makes the United States weak in four main ways. First, the United States becomes vulnerable to disruptions in supply, whether accidental or intentional. Second, dependency entails a massive shift of economic resources from the US to foreign suppliers. Third, dependence requires the US to grant various favors to the leaders of its major suppliers. Foreign suppliers, for example, may make demands for support at the United Nations; these states may also demand access to advanced weaponry or the provision of military protection. Fourth, dependency can jeopardize the security of the United States through entangled foreign alliances or by arousing the hostility of political and religious groups that resent both US military presence and economic ties.[144]

A final argument related to oil is associated with international trade. Currently, oil transactions are conducted in US dollars—a legacy of the shift from a 'gold' standard to an 'oil' standard. Briefly, in 1944 the Bretton Woods Conference established the World Bank and the International Monetary Fund. The Conference also chose the American dollar as the backbone of international exchange, backed by the gold standard, set at $35 an ounce. The Bretton Woods Conference, consequently, was instrumental in the rebuilding of Western Europe and Japan after the Second World War.

During the 1950s and into the 1960s, the system worked well for the United States. However, with the escalating costs of conducting the Vietnam War—coupled with the growing economic strength of Western Europe and Japan—the dollar was weakening. European states became concerned about the dollar's value and began redeeming their dollars for gold. This foretold a potentially devastating drain on the Federal Reserve's gold stocks. In response Nixon unilaterally abandoned the dollar-gold link, thereby establishing a system of floating currencies. However, a free-floating dollar, coupled with a growing US trade deficit and massive debt incurred from the Vietnam War further threatened the US economy. Key members of the Nixon administration, in particular Henry Kissinger, entered into talks with Saudi Arabia to unilaterally price international oil sales in dollars. This process would become known as petrodollar recycling.[145]

---

143 Klare, *Blood and Oil*, 10-1.
144 Klare, *Blood and Oil*, 11.
145 Clark, *Petrodollar Warfare*, 19-20.

During the spring of 1973 a group of 84 of the world's top financial and political insiders met at Saltsjöbaden, Sweden. The purpose of the meeting was to plan to manage the expected windfall that would be generated from petrodollar recycling. Known as the Bilderberg group, participants included Robert Anderson of Atlantic Richfield Oil Co., Lord Greenhill, chairperson of British Petroleum, Sir Eric Roll of S.G. Warburg, George Ball of Lehman Brothers investment bank, David Rockefeller of Chase Manhattan Bank, and Zbigniew Brzezinski, who would soon serve as President Jimmy Carter's national security advisor. Henry Kissinger was also a regular participant at the meetings.[146] William Engdahl describes the objectives of the group:

> What the powerful men grouped around Bilderberg had evidently decided that May was to launch a colossal assault against industrial growth in the world, in order to tilt the balance of power back to the advantage of Anglo-American financial interests and the dollar. In order to do this, they determined to use their most prized weapon—control of the world's oil flows. Bilderberg policy was to trigger a global oil embargo, in order to force a dramatic increase in world oil prices.[147]

A shift from gold-backed currency to an oil-backed currency occurred in the 1970s. By removing the dollar's redemptive value from a fixed amount of gold, the federal government was unfettered in its ability to print new dollars. The only limit was how many dollars the rest of the world would be willing to accept. As Clark details, "if oil can be purchased on the international market only with US dollars, the demand and liquidity value will be solidified, given that oil is the essential natural resource for every industrialized nation." Consequently, US and UK banking elites maneuvered that the Organization of Petroleum Exporting Countries (OPEC) agree to price and conduct all of its oil transactions in dollars. Surplus petrodollars would be used to reverse the dollar's falling international value via high oil prices.[148]

Given that oil must be purchased with dollars, it became imperative for oil-importing states (including Japan) to export cheap goods to the United States in order to gain dollars. These dollars could then be used to buy oil from oil-producing states, such as Saudi Arabia. In turn, Saudi and others invest these dollars into the American economy. This helps keep the dollar strong.

The oil shocks of 1973-1974 were artificially created through the secret negotiations and political manueverings of Kissenger and other members of the Bilderberg Group.[149] Rapid and massive oil price hikes—on the order of 400 percent—created substantial demands for dollars. Furthermore, the oil crisis was compounded by the decision of large multinational oil companies, led by Exxon, to create a short supply of domestic crude oil.

---

146 Engdahl, *Century of War*, 130-4.
147 Engdahl, *Century of War*, 135.
148 Clark, *Petrodollar Warfare*, 29-30.
149 See Engdahl, *Century of Warfare*, 135-8.

In the Third World, oil-importing countries were faced with the challenge of having to acquire export-based dollars to pay for expensive oil-import bills. The vast majority of the world's less developed economies, without significant domestic oil resources, were suddenly confronted with an unexpected and unpayable 400 per cent increase in the cost of energy imports. Sudan, Pakistan, the Philippines, Thailand, and other countries throughout African and Latin America were faced in 1974 with mounting deficits in their balance of payments.[150] Many developing states, including the Philippines, became mired in debt. The US-dominated IMF, however, enforced debt repayments to banks and enforced draconian structural adjustment programs. Consequently, public spending for health, education, and welfare programs decreased substantially. These practices would correspondingly contribute to increased levels of poverty and exploitation throughout developing countries—conditions that would eventually generate support for terrorist organizations. In the United States, however, dollar/petrodollar supremacy allows the government a unique ability to sustain yearly current account deficits, pass huge tax cuts, build a massive military empire of bases worldwide, and still have others accept its currency as a medium of exchange for their imported goods and services.[151]

The oil shock was devastating to world industrial growth, but was extremely beneficial for certain interests, including the major New York and London banks, and the Seven Sisters oil multinationals of the United States and Britain. Indeed, as Engdahl notes, "Chase Manhattan, Citibank, Manufacturers Hanover, Bank of America, Barclays, Lloyds, Midland Bank—all enjoyed the windfall profits of the oil crisis."[152] The rapid rise in the price of oil flooded OPEC members with dollars that exceeded domestic investment needs. These surplus dollars, consequently, were invested in US and UK banks and then re-loaned to governments of developing countries that were desperate to borrow dollars to finance oil imports.[153]

The petrodollar system demands the buildup of huge trade surpluses in order to accumulate dollar surpluses. This is the case for every country—except the United States, which controls the dollar and prints it at will or fiat. Because the majority of all international trade—figured at 70 percent—is conducted in dollars, other governments must engage in active trade relations with the US to obtain the means of payment they themselves cannot issue.[154] The central banks of Japan, China, South Korea, and numerous others all buy US Treasury securities with their dollars; this allows the US to have a stable dollar, far lower interest rates, and a $500-$600 billion annual balance of payment deficit with the rest of the world.[155]

---

150 Engdahl, *Century of Warfare*, 140.

151 Clark, *Petrodollar Warfare*, 28.

152 Engdahl, *Century of Warfare*, 140-1.

153 Clark, *Petrodollar Warfare*, 22.

154 Clark, *Petrodollar Warfare*, 32.

155 Clark, *Petrodollar Warfare*, 33. The United States government, for its part, must ensure that Americans continue to consume these goods. Consumption is increasingly difficult, however, as multinational corporations have re-located manufacturing jobs overseas to take advantage of cheap labor and lax labor and environmental regulations. Most US investment

Until 2000 no OPEC country violated the petrodollar oil price arrangement. However, in September of that year Saddam Hussein proclaimed that Iraq would transition its oil export transactions to the euro currency. A euro-based bank account with BNP Paribas, the leading French bank, was opened and Iraqi oil proceeds went into a special UN account for the Oil for Food program and then deposited in BNP Paribas.[156] Clark identifies the problem: "Although this little-noted Iraqi move to defy the dollar in favor of the euro, in itself, did not have a huge impact, the ramifications regarding further OPEC momentum toward a petroeuro were quite profound. If invoicing oil in euros were to spread, especially against an already weak dollar, it could create a panic sell-off of dollars by foreign central banks and OPEC oil producers." Clark concludes that the prewar diplomatic conflicts and the subsequent reluctance of many European states to support the war are indications of an unspoken war between the dollar and the euro. The Iraq War, consequently, was not about Saddam Hussein's discarded WMD program, the war on terror, or even some altruistic concern over the Iraqi people. Instead, the war was, in part, a forceful and violent message to OPEC and other oil producers to retain the oil-backed dollar.[157]

Access to oil, dependency on foreign sources of the commodity, and the control of the terms of trade all contributed to the invasion of Iraq. However, just as the war in Iraq was not about liberation, neither was it reducible solely to oil. A particularly salient point is raised by Boal and his co-authors: "In the long march toward the modern world system, *mass* commodities—gold, sugar, slaves, cotton, coal, oil—have been its beasts of burden."[158] To reduce the invasion and occupation of Iraq to a war of oil, therefore, misses the larger business of war. Instead, I agree with the interpretation of Boal and his colleagues that the invasion and occupation of Iraq must ultimately be located "in the deadly alchemy of permanent war, capitalist accumulation, and the new enclosures."[159] The 'Blood for Oil' thesis certainly carries

---

in sustainable markets such as manufacturing has become foreign rather than domestic. The American economy is faced with a major disparity between economic growth and the lack of job creation. It is not surprising that Bush implored the American public to consume even more in the aftermath of the September 11 attacks. Speaking to airline employees at Chicago's O'Hare International Airport on September 27, 2001, Bush spoke about the fear generated by the attacks. However, Bush also explained that "one of the great goals of this nation's war is to restore public confidence in the airline industry. It's to tell the traveling public: Get on Board. Do your business around the country. Fly and enjoy America's great destination spots. Get down to Disney World in Florida. Take your families and enjoy life, the way we want it to be enjoyed." As the ruins of the World Trade Center and the Pentagon still smoldered, and efforts continued to recover the remains of victims of the attacks, Bush wanted the American public to 'enjoy life' at Disney World. The War on Terrorism would be won through American consumption.

156 Clark, *Petrodollar Warfare*, 28-31.
157 Clark, *Petrodollar Warfare*, 32, 38.
158 Boal et al., *Afflicted Powers*, 39.
159 Boal et al., *Afflicted Powers*, 43.

weight; however, an over-emphasis on oil mis-describes what a single commodity can actually represent in relation to the larger structural imperatives of the capitalist system.[160]

Orwell's War

Iraq, in the grand strategy of American foreign policy, was simply a pawn. And with all pawns, their use is of limited duration for a particular purpose. Bush, in a sense, was correct when he declared that the 'battle' for Iraq was part of a larger war. This war, however, was not against terrorism but rather against foreign competitors to American hegemony. It was about neoliberalism and global domination; it was part of an effort to secure American military and economic supremacy on a global scale. The Bush Doctrine constitutes a practice of militant neoliberalism, a return to the harshness of previous colonial regimes that facilitated capitalist expansion through brute force. Such is the business of war.

Smith bluntly states that "The new US pugilism, inspired by a political subclass with their social base in the energy/military sector of the capitalist class, can be seen as a strategy for enhanced US control of global oil supply over the next two or three decades vis-á-vis competitors. To the extent that such a strategy succeeded, the US would enjoy unparalleled global economic hegemony."[161] And it is on this score that the moves played by the Bush administration conformed with those of prior administration, and especially that of Clinton. America's involvement in—and sacrifice of—Iraq is hardly unique in US history. As Smith concludes, "Whatever the tactical discrepancies between the Clinton and Bush administrations, neoliberals and neoconservatives, they share entirely the larger goal of an American globalism."[162]

---

160 Boal et al., *Afflicted Powers*, 52.
161 Smith, *Endgame*, 24.
162 Smith, *Endgame*, 25.

Chapter 3

The Business of Occupation

America's war in Iraq was not about Iraq. It was, rather, about America's on-going project of empire building and the remaking of global spaces conducive to capitalist expansion. More specifically, the invasion and subsequent occupation was to gain access to oil reserves; to control the global distribution of oil; to secure the exchange of petroleum in dollars as opposed to euros; to prevent the rise of competing superpowers (especially China) to challenge American supremacy and hegemony; and to open the Middle East to neoliberal capitalism. In essence, the war was a business opportunity, one that was capitalized on by a selective few transnational corporations and multinational banks.

The 'market' though has not determined the course of US foreign policy; neither has the 'market' dictated the economic plunder of Iraq. The Bush administration, following previous administrations, pursued an aggressive neoliberal and neoconservative agenda that has, at its core, a series of basic premises: the preeminence of free-market principles; reductions in government regulations of the economy; free reign for American and multinational corporations; the suppression of trade unions and labor movements; and the elimination of social welfare programs.

The Bush administration has consistently defined and redefined the War on Terror and the subsequent military conflict in Iraq to suit a neoliberal and neoconservative agenda. However, as the writings of Bob Woodward make clear, war was not a matter of last resort for the Bush administration. The primary motivation was economic; to benefit a few select corporations. It was (and is) a class-war, paid for by the lives of soldiers, reservists, and guards who will not 'gain' from the neoliberal exploitation of Iraq. In this respect, the military intervention in Iraq was not unlike numerous other campaigns, such as the war in Vietnam, launched by the United States on foreign territories.

The invasion and subsequent occupation of Iraq reveals the confluence of many transnational practices, although two in particular stand out: the use and movement of private military corporations (warriors) and the deployment of private contract laborers (workers). The increased activities of private military firms (PMFs) and overseas contract labor migration are intimately connected, both seen as part of a military neoliberalism. Combined, the transnational movement of warriors and workers demonstrates the business of war and why Occupied Iraq signifies a neoliberal, privatized and ultimately de-humanized space. Both workers and warriors are commodified entities, deployed to facilitate capital largess.

According to Ian Traynor, by December 2003 private military contractors constituted the second biggest contributor to coalition forces in Iraq, surpassed only by US military personnel. Private warriors, in other words, outnumbered the total troop commitment of the United Kingdom. Moreover, Traynor identifies that of US$87 billion earmarked in 2003 for the Iraqi War, one-third of that, nearly US$30 billion was to be spent on contracts to private companies. He concludes that "The private sector is so firmly embedded in combat, occupation and peacekeeping duties that the phenomenon may have reached the point of no return: the US military would struggle to wage war without it."[1]

In this chapter I discuss the confluence of warriors and workers as they coalesce in Occupied Iraq, for it is this connection that sets the spaces for the political subjugation of hostages. Special attention is directed toward the Philippines, in part because this country is the largest supplier of contracted labor in Occupied Iraq, servicing US military bases and other construction sites throughout the country. As such, I perceive Philippine overseas employment as the counter-part to the militarism of the United States (discussed in Chapter 2).

Setting the Occupation

On 20 March 2003 the ground invasion of Iraq began. By 5 April US forces entered Baghdad and within days the capital city fell. The downfall of Saddam Hussein's brutal dictatorial regime was rapid and, to most observers, not unexpected. What remained unclear, however, was the aftermath. There exists scepticism, for example, as to the efficacy of pre-war planning for a post-war Iraq. According to David Phillips, the Bush administration never had a solid or viable plan or a program for running postwar Iraq.[2] Phillips, who was a member of the Future of Iraq Project, maintains that despite masses of information on Iraq and what may lay ahead, no concerted effort was made to incorporate such knowledge. The Future of Iraq Project, as a case in point, was established in late 2002 and entailed a series of conferences, drafting sessions, seventeen working groups on post-war Iraq. None of these materials, however, were incorporated in the actual reconstruction of post-war Iraq and the CPA. To Phillips, "The Bush administration had goals for Iraq, but no coherent strategy for accomplishing them. Its policy was based on a combination of naivete, misjudgement, and wishful thinking."[3]

Larry Diamond concurs, noting that although the United States invaded Iraq without a coherent, viable plan to win the peace, the absence of a plan does not equate with 'no preparation'. Diamond, who worked in Iraq from January to April 2004 at the request of then-National Security Advisor Condoleezza Rice, describes the lack of preparation for post-war as one of gross or criminal negligence. He writes: "As

---

1   Ian Traynor, "The Privatization of War," *The Guardian/UK*, December 10, 2003, www.commondreams.org/cgi-bin/print.cgi?file=/headlines03/1210-13.htm (September 9, 2005).

2   David L. Phillips, *Losing Iraq: Inside the Postwar Reconstruction Fiasco* (New York: Westview Press, 2005), 67.

3   Phillips, *Losing Iraq*, 156.

Coalition soldiers and civilians were being attacked and killed in growing numbers, we still did not have enough armored cars, trucks, and Humvees, and enough high-quality body armor. Everyone knew we faced crippling shortages, but no one in authority viewed the situation as urgent."[4] Such occurrences did not seem to affect members of the Bush administration. Indeed, it was Secretary of Defense Donald Rumsfeld who disingenuously remarked that "You go to war with the army you have."[5]

But this is entirely the point. There were in fact various plans, interpretations, and agendas. Perhaps, ironically, too many. The Future of Iraq Project, for example, spent $5 million but was too ambitious. Headed by the State Department, the project developed recommendations including education and agriculture, health and sanitation, energy and public finance.[6] Consequently, it was later canceled. A second problem stemmed from the rush to war. Little concerted thought was given the details of occupation in part because those who were tasked with the occupation had little time to do so.

By March 2002 regime change in Iraq had become the main objective of the Bush administration. Irregardless of the discussions that echoed throughout the United Nations, there was—for most members of the Bush administration—no option other than war. However, despite the near unanimity of the Bush administration in overthrowing the Saddam Hussein regime, different opinions existed as to the end-result and overall purpose of the campaign. Vice President Richard Cheney, arguably the most vociferous in his desire to topple Saddam, saw the war in Iraq as a means to project US military power. Rumsfeld viewed the conflict in similar terms. Deputy Defense Secretary Paul Wolfowitz and Deputy Secretary of State Richard Perle both envisioned Iraq as a prototype for spreading democracy throughout the Middle East. Iraq, it was assumed, would have a ripple effect, causing regime change in Syria and Iran. Wolfowitz, moreover, who had served briefly as ambassador to Indonesia and witnessed first hand the 'peaceful' overthrow of former Philippine president and dictator Ferdinand Marcos, saw in Iraq an opportunity to establish a secular, pro-Western anchor for American interests in the Muslim world. Perle, a right-wing hawk, supported the attack in a belief that an overwhelming US victory would send a message to other, less friendly regimes.[7] Lastly, Colin Powell, as Secretary of State, and General Jay Garner, who would serve as the first US envoy to post-war Iraq, were among the few pragmatists who actually worked to prevent war. Powell advocated a greater reliance upon diplomacy; global threats were to be met through

---

4   Larry Diamond, *Squandered Victory: The American Occupation and the Bungled Effort to Bring Democracy to Iraq* (New York: Henry Holt and Company, 2005), 292-3.

5   *BBC News*, "Troops Grill Rumsfeld Over Iraq," *BBC News*, 8 December 2004, http://news.bbc.co.uk/2/hi/middle_east/4079201.stm (20 February 2006).

6   Phillips, *Losing Iraq*, 37.

7   David Phillips, *Losing Iraq: Inside the Postwar Reconstruction Fiasco* (New York: Westview Press, 2005), 56-60.

international cooperation. It is no surprise that Powell was the only senior national
security advisor who had seen combat and had served in the Vietnam war.[8]

Although plans for war were well on their way by late 2002, it was not until
20 January 2003 that the Bush administration established the Orwellian-named
Office of Reconstruction and Humanitarian Assistance (ORHA). The ORHA was
headed by retired Army Lieutenant General Jay M. Garner, hand-picked by long-
time friend Rumsfeld. Prior to his appointment as head of the ORHA, Garner served
as president of SY Coleman, a defense contractor specializing in missile-defense
systems, including the Patriot missile. Garner, who had virtually no private-sector
experience, was named president of the firm in 1997 after serving as vice chief of
staff of the Army.[9]

The appointment of Garner would foreshadow a process that would come to
define the occupation of Iraq, namely the prevalence of cronyism and favoritism.
According to David Kirp, a professor at the University of California at Berkeley's
Goldman School of Public Policy, the Bush administration sent a profound message
to the Iraqi people by placing Garner in charge of reconstruction and humanitarian
aid. Kirp explained in an interview prior to the invasion that "This is a lovely
example of our indifference to the people of Iraq. It truly bespeaks a lack of serious
thinking on the administration's part."[10] David Armstrong, a defense analyst for the
Washington-based National Security New Service, agreed, noting that "It seems
inappropriate for somebody to step into a humanitarian and administrative role from
a company with a role in providing equipment which, albeit defensive, is vital to the
success of the US operation."[11]

Garner was ill-prepared for the task at hand. He lacked of regional expertise and
his views on post-war Iraq were later revealed to be inconsistent with the majority
opinion of the Bush administration. And the invasion of Iraq would begin in less
than two months, a time-schedule that did not leave much time for the ORHA to
make adequate preparations. Time did permit, however, the doling out of no-bid
and questionable bid contracts. Two months before the war began USAID awarded
Bearing Point, an offshoot of the accounting firm KPMG, to oversee Iraq's transition
to a 'sustainable market-driven economic system.'[12] According to Pratap Chatterjee,
Bearing Point initiated a three-year privatization process that was to include a
comprehensive income tax system, a new banking system, and a rapid and thorough
exchange of currencies. Subsequently, Bearing Point hired the New Jersey-based

---

8   Phillips, *Losing Iraq*, 56-7.

9   David Lazarus, "General Reverses his Role," *San Francisco Chronicle*, February
26, 2003, http://sfgate.com/cgi-bin/article.cgi?file=/chronicle/archive/2003/02/26/BU48310.
DTL&ty (September 10, 2005).

10   Lazarus, "General Reverses."

11   Oliver Morgan, "US Arms Trader to Run Iraq," *Guardian Unlimited*, March 30, 2003,
http://observer.guardian.co.uk/business/story/0,6903,925309,00.html (September 10, 2005).

12   Naomi Klein, "Baghdad Year Zero: Pillaging Iraq in Pursuit of a Neocon Utopia,"
Harper's Magazine, September 2004, http://harpers.org/BaghdadYearZero.html (September
9, 2005).

engineering firm Louis Berger Group as a subcontractor to oversee the collection and destruction of old Iraqi currency and the introduction of a new one, which was completed before the end of spring 2004.[13]

On 11 March, in a background briefing on the reconstruction of Iraq, a senior defense official explained that following Bush's directive of 20 January establishing the ORHA, the organization started very slowly. During its first week of existence the ORHA had only about three people, though it had grown to close to 200 by March. The official was hesitant, almost unsure of how to proceed: "We've taken all the plans—well, not all—all the plans that we know of, that have been prepared by the interagency—and there's been a—there was an awful lot of work done by the interagency. We brought all those together and we read them—we haven't changed any—we've read them, and we began trying to connect the dots on all of them." The goal, according to the official, was to "put together a solid set of plans" to be implemented, and for the United States to stay only as long as necessary before turning Iraq over to the Iraqi people. Indeed, the official stated that "Our time frame in country is to get in there as soon as we can and begin this work, and end it as fast as possible."[14] Bush, Rice, Wolfowitz continued to reiterate that Iraqi interim government would be established as soon as possible.

On 6 April, the date after US ground forces entered Baghdad, it was reported that National Security Advisor Condoleezza Rice ruled out any key role for the United Nations in the reconstruction of Iraq. This announcement was part of a larger political maneuver whereby the United States attempted to establish its own unilateral and hegemonic rule over Iraq. Even loyal Coalition members, such as the United Kingdom, risked being shut out by American officials.[15] This attempt, however, was met with resistence from key members within the Bush administration. Powell continued to favor a military government that would facilitate a rapid transition to Iraqi governance; Garner likewise favored a short occupation period, three months at most.[16]

Pentagon officials, on 14 April, announced the cessation of major military operations. Garner entered Iraq one week later. His arrival in Baghdad was met by an anti-US demonstration of approximately 2,000 Shiite Muslims shouting "no, no to colonialism." Garner said in an interview: "What better day in your life can you have than to be able to help somebody else, to help other people, and that is what we intend to do." When touring Baghdad's 1,000 bed Yarmuk hospital, Garner

13 Pratap Chatterjee, *Iraq, Inc.: A Profitable Occupation* (New York: Seven Stories Press, 2004), 178.

14 United States Department of Defense, "Backgrounder on Reconstruction and Humanitarian Assistance in Post-War Iraq," March 11, 2003, www.defenselink.mil/transcripts/2003/t03122003_t0311bgd.html (September 8, 2005).

15 Ed Vulliamy and Kamal Ahmed, "US Begins the Process of 'Regime Change,'" *Guardian Unlimited*, April 6, 2003, http:observer.guardian.co.uk/international/story/0,6903,930794,00.html (September 8, 2005).

16 Vulliamy and Ahmed, "US Begins."

explained that "We will help you but it is going to take time." An Iraqi doctor noted that "If they give us anything it is not from their own pockets. It is from our oil."[17]

The comments of the Iraqi doctor speak volumes. The Iraqi citizenry understood what many in the United States did not: the American venture into Iraq was a unilateral exercise of capitalist expansion, predicated on the suppression of true democracy or liberation. The prime directive of occupation forces was to secure and re-establish the oil infrastructure of Iraq. Ostensibly, revenue from oil production would help re-build the country. However, it is questionable as to whether this would in fact materialize. And regardless of the long-term distribution of oil revenues, Iraqi citizens were suffering from a lack of basic supplies. Devastated by years of economic sanctions, and compounded by the 'shock and awe' of 'liberation,' the Iraqi infrastructure was inoperable. Water and electricity, sanitation and sewage: these demanded immediate attention. That these basic services were not addressed first highlights the hollowness of humanitarianism as a justification for war.

During the brief weeks in which Garner was in charge of Iraq's reconstruction, it became apparent that his vision for a post-war occupation was diametrically opposite that of the neo-conservatives who pressed for war. Speaking before the BBC on 15 April, Garner explained that "You don't try to build [a country] in the image of your own country. You open it to the people and you begin a dialogue with the people and let them begin a dialogue with themselves."[18] Officials such as Wolfowitz and Rumseld, conversely, favored a neocolonial situation, one whereby the United States would impose its own values and unilaterally open the country to foreign investment. Dialogue, especially between 'the people,' would jeopardize these wider goals.

Naomi Klein maintains that for the neo-conservatives, Garner's approach to post-war Iraq was hopelessly unambitious. Instead, they saw in Iraq an opportunity to open the country to become "a model corporate state that would open up the entire region."[19] In conforming with neoliberal ideology, occupation was anticipated to be a lean operation, relying on private contractors to accomplish most of the reconstruction effort. Such an arrangement, moreover, would be more lucrative to selected US companies such as Halliburton and Bechtel.

The Coalition Provisional Authority

On 1 May 2003 the Bush administration announced that L. Paul Bremer III would oversee the selection of the Iraqi transitional government. Five days later Bremer was named by Bush as new civilian administrator of postwar Iraq and would be the head of the newly-formed Coalition Provisional Authority (CPA). In this position, Bremer would exercise all executive, legislative, and judicial power, manage all

17 "Jay Garner Tours Baghdad," *Guardian Unlimited*, April 21, 2003, www.guardian. co.uk/Iraq/Story/0,2763,940683,00.html (September 9, 2005).

18 "General Jay Garner Speaks to BBC," *BBC News*, April 15, 2003, http://news.bbc. co.uk/2/hi/middle_east/2948907.stm (September 9, 2005).

19 Klein, "Baghdad Year Zero."

ministries, and supervise the drafting of a constitution. In the process, however, the idea of an interim government with real sovereign authority had been indefinitely postponed.[20] Bremer and the CPA arrived in Baghdad on 11 May 2003. He would remain in Iraq for thirteen months until the transition to Iraqi interim government of 28 June 2004.

Throughout his career, Bremer has served under numerous presidential administrations in a variety of posts. During a twenty-three year diplomatic career, Bremer was stationed in Afghanistan, Malawi, and Norway; he also served as Ambassador to the Netherlands. In 1989 he worked for the New York-based Kissinger Associates and, in 2001, co-chaired (with former Attorney General Edwin Meese) the Heritage Foundation's Homeland Security Task Force. In June 2002 Bush appointed Bremer to the President's Homeland Security Advisory Council.[21]

Recognized as an expert in terrorism, counter-terrorism, and homeland security, Bremer has maintained a hard-line against terrorists. He advised the Clinton administration to deliver ultimatums against Libya, Syria, Iran, and the Sudan— ironically, Iraq was not among his list. Bremer has also recognized, however, that terrorist threats represent lucrative business opportunities. On 11 October 2001—just one month following the attacks on the World Trade Center and the Pentagon—Bremer joined with Jeffrey Greenberg, chair and chief executive officer of Marsh & McLennan Companies to launch a new consulting unit to capitalize on corporate fears of terrorism. The newly formed March Crisis Consulting would sell insurance to corporate customers at sharply higher rates than were common prior to 11 September 2001.[22] During a panel discussion on terrorism risks at the Risk & Insurance Management Society Inc.'s annual conference, held in New Orleans during April 2002, Bremer was asked about the possibility of another '9/11-type' attack in the United States. Bremer replied, "I think there is a 100% chance we will see another attack of the same severity or worse; we just don't know the timeframe."[23] No doubt the pronouncements of a recognized terrorist expert such as Bremer would profit Marsh Crisis Consulting. It was this ethically-suspect form of cooperation between government and the private sector that would characterize the occupation of Iraq. Bremer, as the embodiment of 'fear-for-profit' schemes, was ideologically in line with the neo-conservatives of the Bush administration, namely Paul Wolfowitz and Richard Perle.[24]

Bremer approached the reconstruction of Iraq with the same anti-intellectualism, favoritism, cronyism, and incompetency that exemplified the post-war planning efforts. Donald Rumsfeld, for example, opposed any role for career State Department

---

20 Diamond, *Squandered Victory*, 37.

21 Bill Berkowitz, "Bremer of Iraq," May 9, 2003, www.alernet.org/module/printversion/15864 (September 14, 2005).

22 Berkowitz, "Bremer of Iraq."

23 Sarah Veysey, "Terrorism Heightens Need for Crisis Plans," *Business Insurance*, April 29, 2002, www.businessinsurance.com/cgi-bin/article.pl?articleID=9473&a=a&abt=ABS&arc=n (September 14, 2005).

24 Phillips, *Losing Iraq*, 144.

experts.[25] Of the 1,147 Americans employed by the CPA in July 2003, only 34 were Foreign Service Officers; the CPA lacked both area and language expertise. Likewise Bremer and other members of the CPA refused to cooperate with anyone or any institution that did not conform ideologically with their neoconservative outlook.[26] Klein describes some of those working under Bremer at the CPA:

> Jay Hallen, a twenty-four-year-old who had applied for a job at the White House, was put in charge of launching Baghdad's new stock exchange. Scott Erwin, a twenty-one-year-old former intern to Dick Cheney [and senior at the University of Richmond], reported in an email home that "I am assisting Iraqis in the management of finances and budgeting for the domestic security forces." The college senior's favorite job before this one? "My time as an ice-cream truck driver."[27]

In late June 2003 Bremer outlined his vision for a free-market Iraq before hundreds of executives attending a meeting of the World Economic Forum in Jordan. Bremer explained that "Markets allocate resources much more efficiently than politicians. So our strategic goal in the months ahead is to set in motion policies which will have the effect of reallocating people and resources from state enterprises to more productive private firms."[28] Bremer elaborated that "Getting inefficient state enterprises into private hands is essential for Iraq's economic recovery."[29] As a case in point, on 19 September 2003 Bremer issued CPA Order 37: the 'Tax Strategy for 2003.'[30] This order suspended income and property taxes for all those associated with the occupation, including the CPA; Coalition Forces; Forces of countries, their contractors and sub-contractors, acting in coordination with Coalition Forces; Coalition contractors and sub-contractors [that] supplied goods directly to or on behalf of the CPA and Coalition Forces; Departments and agencies of Coalition Forces' governments, and their contractors and sub-contractors that are providing technical, material, financial and human resource assistance to Iraq; Government, international organizations, and not-for-profit organizations providing technical, material, financial and human resource assistance to Iraq; and members of the preceding categories to whom goods are consigned or by whom goods are imported for their personal use. In short, Iraqis would largely be the only ones paying taxes in the country.

This date also saw the issuance of CPA Order 39 which permitted foreign investors to fully own Iraqi companies with no requirements for reinvesting profits back into

---

25 Diamond, *Squandered Victory*, 30.

26 Phillips, *Losing Iraq*, 163.

27 Klein, "Baghdad Year Zero"; Erwin was shot in an ambush while working in Iraq, two others died in the attack. Erwin returned to Richmond in 2004 and was subsequently awarded the Defense of Freedom Award by the Pentagon.

28 Chatterjee, *Iraq, Inc.*, 178.

29 Klein, "Baghdad Year Zero."

30 "Coalition Provisional Authority Order Number 37: Tax Strategy for 2003," September 21, 2003, www.cpa-iraq-org/regulations/20030921_CPAORD.pdf. (September 9, 2005).

the country. Profits, moreover, would be exempt from taxation. The only industry exempted from privatization was the oil industry, which was to remain under the guidance of American officials. Moreover, under Executive Order 13303, signed by Bush in late May 2004, Iraqi oil contractors were given a lifetime exemption from lawsuits. Consequently, multinational oil companies are immune from legal proceedings in the event of any accident or wrong-doing: tanker accidents, explosions at oil refineries, employment of slave labor, lawsuits by governments demanding compensation. All actions are immune from judicial accountability.[31]

The privatization of Iraq required considerable political maneuvering. Internationally, the Bremer regime treated Iraq as though it were a prize to keep, and refused to allow other states—even Coalition members—any real decision-making authority in Iraq. Internal to Iraq, the Bremer regime likewise set in place a classic neocolonial relationship, complete with puppet governments. On 13 July 2003, for example, Bremer appointed the twenty-five member Iraqi Governing Council. Previously, Garner had pledged to establish an interim Iraqi administration; the arrival of Bremer reversed this course of action. Bremer instead announced an advisory committee that would have only limited powers; the handover of sovereignty was post-posed indefinitely.

The advisory council was "selected, not elected" by the Coalition Provisional Authority. According to Phillips, the CPA hand-picked local leaders to participate in neighborhood councils; these councils, in turn, would select district councils, which would finally select governors. All decisions had to be approved by US military commanders, a situation that exacerbated the Iraqis' resentment.[32] In response to strong objections by Iraqi leaders, Bremer renamed the advisory council the Iraqi Governing Council (IGC). The IGC was given only limited authority. The Council, for example, did not have authority over the basing of foreign troops in Iraq, Iraq's oil sector, or foreign affairs.[33] Consequently, it appeared as if Bremer would have free reign in the privatization of Iraq.

Prior to the invasion, Iraq's non-oil related economy was dominated by 200 state-owned companies, which produced everything from cement to paper to washing machines.[34] Thomas Foley was to oversee the privatization of the Iraqi economy. Foley, a multimillionaire, major fund-raiser, and former chair of Bush's Connecticut campaign finance committee in 2000, was appointed in August 2003 and charged to run the Iraqi state business sector. Also on board was Michael Fleischer, a venture capitalist and brother of former White House spokesperson Ari Fleischer. In October Foley indicated that he would present the recently formed Iraqi Government Council with a proposal on privatizing state-owned businesses within seven months. According to Chatterjee, Foley's was to sell off 150 of the 200 state-owned enterprises—but exclude electricity assets and financial institutions, as well as the oil industry. By

---

31 Chatterjee, *Iraq, Inc.*, 179.
32 Phillips, *Losing Iraq*, 170.
33 Phillips, *Losing Iraq*, 172.
34 Klein, "Baghdad Year Zero."

February 2004, however, Foley conceded that his proposal was failing. The IGC in fact had completely rejected his plan for privatization, and Foley left Iraq by the end of the month.[35] Although this latest bid was unsuccessful, it reveals the colonial intent of the occupying forces and contributed to the mounting unrest that would culminate in the violent insurgency directed against Coalition forces.

Peopling the Insurgency

One of Bremer's first actions was to 'de-Ba'athify' Iraq. On 16 May 2003 Bremer decreed that all senior party members would be removed from civil and military positions, and that they would be banned from future employment. Overnight Bremer effectively fired 120,000 Iraqis—including teachers and doctors. It was believed, after dismantling the Ba'ath Party, that Iraq's technocrats would transfer their loyalties to a new administration and that Iraq would continue to function more or less as before.[36] The Bush administration condoned these actions.[37]

This was a case of guilt by association. The administration did not distinguish between those who loyally supported the Ba'ath government, and those who did so for personal and familial safety reasons. It was illogical to believe or to presume that all 120,000 fired Iraqis were guilty of committing atrocities. Of course, the dismissal of party members was not solely about guilt. On the one hand, the Bush administration sought to distribute power to Iraqis who subscribed to its vision of the new Iraq.[38] In such a move, the cronyism of the Bush administration was distributed to higher levels within Iraq. On the other hand, the purge of Ba'ath Party members also eliminated potential opponents to the liberalization of Iraq's national economy. Bremer and other neo-conservative policy-makers believed that privatization would open Iraq to foreign investment, bring in foreign companies, and create an Iraqi bourgeoisie whose prosperity would inspire Syrians, Iranians, and others to seek the same.[39]

One week later, on 23 May, Bremer issued an edict targeting the armed forces of Iraq. He issued a list of disbanded entities that included all branches of the armed forces, intelligence agencies, and paramilitary groups (e.g., the Saddam Fedayeen). Moreover, all Iraqis holding the rank of colonel or higher were considered to be 'senior' party members and were subsequently denied payment. Other personnel were offered a termination payment. Benefits to widows of deceased senior party members were eliminated. Bremer also announced the establishment of the New Iraqi Civil Defense Corps as a first step in forming a national defense capability.[40]

---

35 Chatterjee, *Iraq, Inc.*, 182.
36 Phillips, *Losing Iraq*, 8.
37 Phillips, *Losing Iraq*, 145.
38 Phillips, *Losing Iraq*, 146.
39 Phillips, *Losing Iraq*, 147.
40 Phillips, *Losing Iraq*, 149.

Strangely, the disbanding of the armed forces was contradictory to American policies during the war, a contradiction that added to the unrest brewing in Iraq. During the war, for example, Coalition forces released leaflets encouraging Iraqi soldiers to defect and return home. Their light resistance made the military victory easier. After liberation, these military personnel expected to assume a role in the reconstructed armed forces of Iraq. Instead, Bremer sent a pink slip to 400,000 army personnel.[41] Moreover, according to Phillips, the decision to disband the armed forces "flew in the face of advice from both Iraqis and US experts," including US field commanders. Not only did the decree turn 400,000 former soldiers against the US-led coalition, if one considers the average size of an Iraqi household, the decision directly affected the lives of 2.4 million people—roughly 10 percent of Iraq's population.[42]

Ironically, the Garner-led ORHA had explicitly stated its intention of utilizing the former Iraqi military. According to a senior defense official, the ORHA identified that reconstruction would be a "labor-intensive business." As such, one goal was to use the Iraqi army—but not the Republican Guard—to help in the rebuilding process. "The regular army," this official explained, "has the skill sets to match the work that needs to be done in construction." Members of the former Iraqi army would under this plan have continued to be paid and perform such services as the rebuilding of bridges, the removal of rubble, and clearing of unexploded ordinance. This official concluded that by *not* demobilizing the army, the ORHA would not "put a lot of unemployed people on the street." Rather, they would be "working to rebuild their country."[43] Such a proposal, however, would have prevented the awarding of lucrative contracts and sub-contracts to loyal members of the Coalition. For the Bremer group, business took precedence over reconstruction.

Phillips concludes that "The Bush administration had committed one of the greatest errors in the history of US warfare: It unnecessarily increased the ranks of its enemies. Embittered Arab Sunnis, who had dominated the military establishment, would reemerge to lead the insurgency against US troops in the Sunni triangle. The mistakes of disbanding the armed forces and concerning de-Ba'athification would have lasting and far-reaching ramifications."[44] Diamond agrees, writing that these decisions were "criticized at the time by many Iraqis and outside experts who predicted that expelling so many people so suddenly from the postwar order would generate a severe backlash."[45]

This raises another angle to the insurgency, one that I will discuss in greater detail in the next chapter. While foreign workers and contractors were introduced into Occupied Iraq, a staggering 65 percent of Iraqis were estimated to be unemployed. These Iraqis, not surprisingly, would potentially join the insurgency. However, the

---

41 Phillips, *Losing Iraq*, 149.
42 Phillips, *Losing Iraq*, 151-2.
43 United States Department of Defense, "Backgrounder."
44 Phillips, *Losing Iraq*, 153.
45 Diamond, 39.

introduction of foreigners into Iraq would also significantly affect the perception of the occupation. Foreign workers were not seen as liberators, entering the country to help the Iraqi people, but instead were viewed as taking Iraqi jobs. The out-sourcing of the occupation effort, in the face of massive unemployment made worse by the decisions of Bremer, contributed greatly to the continued violence in the country.

Neocolonial Spaces

Bremer had no intention of relinguishing power to any Iraqi government so soon. He had ambitious plans for rebuilding Iraq that required more than a few months of direct US rule. These plans would including wide-ranging free-market economic reforms, such as tax reductions and subsidies, streamlined regulations, foreign trade, and a reallocation of resources from state control to private enterprise.[46] And far from supporting democracy, the Bremer administration sought to censure the country. On 28 March 2004, for example, Bremer ordered the Iraqi newspaper Al-Hawza shut down for two months. Bremer claimed that the newspaper incited violence against coalition troops. Nevertheless, his decision increased an already dangerous and tense situation, contributing to an anti-coalition uprising across central and southern Iraq. According to Associated Press writer Tarek El-Tablawy, Bremer's decision drew condemnation from members of the now-defunct IGC as well as some officials of Bremer's CPA. Indeed, former CPA officials said in private that the decision had unnecessarily angered a large segment of Iraq's Shiite majority. Interim Prime Minister Iyad Allawi, following the departure of Bremer, would later order the paper to be reopened in an effort to show his "absolute belief in the freedom of the press."[47]

The CPA and, by extension, the Bush administration envisioned a radical neoliberal remaking of Iraq. Consider, for example, the formation of the Iraqi Governing Council (IGC). Established on 13 July 2003, this organization was, in principle, representative of Iraq's demographics. Moreover, the IGC was recognized by UN Security Council Resolution 1500. However, Bremer gave the IGC only limited powers. Foreign nationals, especially Americans, continued to run the government. The IGC did not have the authority to influence the basing of US and coalition troops; Iraq's oil sector, or foreign affairs. Consequently, the IGC operated, and was seen, as a puppet government of the United States.[48]

Bremer and other top American officials also proved unwilling to incorporate the principle players—Ba'athists and Arab nationalists—who would have been capable of defusing the Sunni-based resistance, and who were, in fact, sending signals

---

46 Diamond, 42.

47 Tarke El-Tablawy, "Iraqi Prime Minister Opens Controversial Newspaper Shut Down by Coalition," *San Francisco Chronicle*, July 18, 2004, www.sfgate.com/cgi-bin/article. cgi?file=/news/archive/2004/07/18/international10 (September 14, 2005).

48 Phillips, *Losing Iraq*, 172.

that they wanted to talk directly to the United States.[49] These elements however would have balked at the neoliberal reforms proposed by the CPA. But then, neither democracy nor liberation was the goal. Territorial control over a strategic location in the Middle East, and the subsequent control over resources—oil—was the principle objective.

Sovereignty was allegedly handed to Iraq on 28 June 2004—two days ahead of schedule.[50] An interim government was ostensibly in charge, with elections to be held in January 2005 for a permanent government—twenty months after Bush declared 'mission accomplished.' In the interim, another 1,195 US soldiers and untold numbers of Iraqis would die.

The Illegality of Occupation

Reconstruction efforts targeted selected infrastructure, namely those that would facilitate primarily the occupation forces. Reminiscent of colonial practices, improvements to services and transportation systems did not translate into improvements in the day-to-day lives of the occupied citizenry. The dredging of the Umm Qasr seaport and the Baghdad airport received top priority, as did the oil infrastructure of the country. These projects were targeted because of the military need to bring in equipment for the occupation. Additionally, as Chatterjee notes, within weeks in the spring of 2003, mobile phone towers had sprung up to provide MCI service to American officials and their appointed Iraqi advisors. Meanwhile, hundreds of thousands of Iraqis continue to suffer due to lack of adequate water and electricity. Enduring the searing heat of Iraq's summer, with temperatures in excess of 125 degrees Fahrenheit, locals were in dire straits.[51] In May 2004, over one year after the occupation began, Baghdad's Kerkh sewage treatment plant, which was designed to handle the waste of more than two million people, was still not working. Reports began to surface identifying gross negligence and mismanagement on the part of the contractors, including Bechtel. Consequently, incidents of cholera,

---

49 Diamond, 295.

50 The sudden transfer of authority and Bremer's departure two days ahead of schedule has been subject to considerable speculation. The Bush administration claimed that the early transfer was a positive (and patronizing) sign of American confidence in the Iraqi peoples' ability to govern themselves. However, other sources suggested that Bremer's departure was hastened by rumors of increased attacks to coincide with the publically announced date of June 30. Still others have suggested that Bremer's removal resulted from an alleged love affair between Bremer, a married man with two children, and an Iraqi woman. Whatever the reason, Bremer's sudden departure and the unexpected transfer of authority caused administrative difficulties for those who remained. See "Arab Paper Starts Bremer Rumor," July 1, 2004, www.worldnetdaily.com/news/article.asp?ARTICLE_ID=39236 (September 14, 2005).

51 Chatterjee, *Iraq, Inc.*, 68.

kidney stones, and diarrhea increased throughout the country.[52] This translated into sustained high rates of infant mortality.

Equally if not more egregious is the fact that these transnational practices have all been illegal. As Klein details, the CPA derived its legal authority from UN Security Resolution 1483. Passed in May 2003, this resolution recognized the United States and Britain as Iraq's legitimate occupiers and was to 'promote the welfare of the Iraqi people through the effective administration of the territory.' However, the resolution also stated that the United States and Britain must comply with existing international law, including the Geneva Convention of 1949 and the Hague Regulations of 1907. These earlier conventions had been drafted to prevent occupying powers from economically stripping occupied territories. Occupying forces must abide by a country's existing laws unless 'absolutely prevented' from doing so. Moreover, occupying forces do not own the public buildings, real estate, forests and agricultural assets, but instead is the 'administrator' and custodian, keeping them secure until sovereignty is reestablished.[53] Bremer, however, would cite Resolution 1483 as justification for sidelining the plans to constitute an Iraqi interim government and for exercising governing authority directly.[54] Ironically, the deposed head of the ORHA, Jay Garner, had stated just days before his release that "the Coalition is bound by the Geneva Conventions and the Hague Conventions. We're a very moralistic society and a very honorable society."[55]

The privatization of Iraq's economy as proposed by the CPA was in violation of international law. Consequently, trade lawyers advised their corporate clients to not venture into Iraq just yet, suggesting that it would be better to wait until after authority was transferred to the Iraqis. Otherwise, investments made under Bremer's reign could potentially be expropriated, leaving firms with no recourse. Even US-appointed Iraqi politicians shied away from the ambitious privatization plans, a factor that contributed to the twenty-five member Iraq Governing Council to decide unanimously to not participate in the privatization of Iraq's state-owned companies or of its publically owned infrastructure.[56]

Bremer's guidance of Iraq's reconstruction was characterized by mismanagement, fraud, corruption, and cronyism. The lives of both Iraqis and non-Iraqis, including US military personnel, were sacrificed in an effort to force open Iraq's economy to private gain. Moreover, American tax-payers have been fleeced of their 'investments' as a select few transnational corporations reaped profits. On 30 January 2005, an official report prepared by Stuart Bowen, Special Inspector General for Iraq Reconstruction, concluded that "the CPA did not establish or implement sufficient managerial, financial, and contractual controls to ensure [Development Fund for Iraq (DFI)] funds were used in a transparent manner or that there was no assurance

---

52   Chatterjee, *Iraq, Inc.*, 81.
53   Klein, "Baghdad Year Zero."
54   Diamond, *Squandered Victory*, 38.
55   "General Jay Garner Speaks," *BBC News*.
56   Klein, "Baghdad Year Zero."

the funds were used for the purposes mandated by Resolution 1483." As one case in point, the reported cited that "CPA officials authorized payments of DFI funds for approximately 74,000 Facilities Protective Services (FPS) guards' salaries even though the FPS sites and number of guards were not validated. CPA staff identified at one ministry that although 8,206 guards were on the payroll, only 602 guards could be validated. CPA staff at another ministry validated the payroll at one FPS site and found that although 1,471 guards were on the payroll, only 642 guards could be validated."[57] In short, Bowen found that approximately US$9 billion for the reconstruction of Iraq 'disappeared' through fraud, corruption, and mismanagement. On the one hand, this is not surprising given the high degree of favoritism and nepotism that undergirded the staffing decisions of the CPA. On the other hand, this also bespeaks a more nefarious exploitation of occupied Iraq. Consider the following: most contracts for the reconstruction of Iraq were awarded to well-connected US firms; only two percent of reconstruction contracts in 2003 were awarded to Iraqi firms. Bremer, however, is less than remorseful. Speaking before an audience at Clark University, Bremer was asked about the reports of US$9 billion missing from the funds to rebuild Iraq. Bremer replied that people shouldn't worry inasmuch as the money was Iraqi money and not US money.

Servicing War

War, in the twenty-first century, has increasingly become a business venture. But these ventures are also associated with corresponding migrations and, in particular, the use of overseas contract workers. As privatized military firms (PMFs) circle the globe in search of new opportunities, they must rely on equally flexible supplies of labor to perform their operations. Flexibility and mobility are thus crucial components in the business of war. And different tasks require different types of labor. Armed guards are hired to protect administrators; construction workers are contracted to build bridges and repair facilities. Still more workers are hired to feed and cloth the warriors.

Occupied Iraq has become a corporate space, dominated by transnational corporations and private military firms. Combined, these companies largely dictate the practices underlying the occupation and reconstruction programs in Occupied Iraq. As discussed in Chapter 1, PMFs assume a wide range of duties, including construction efforts and security. It was Global Risk International, for example, an English PMF based in Hampton, Middlesex, that provided protection in the form of Gurkas, Fijian paramilitaries, and ex-SAS veterans for the Baghdad headquarters of Paul Bremer.[58] More notable, though, is the military support firm of KBR. Since 1992, the firm has operated in Afghanistan, Albania, Bosnia, Croatia,

---

57 Stuart W. Bowen, Jr., "Oversight of Funds Provided to Iraqi Ministries through the National Budget Process" (Report No. 05-004), January 20, 2005, www.sigir.mil/pdf/dfi_ministry_report.pdf, (September 14, 2005).

58 Traynor, "Privatization of War."

Greece, Haiti, Hungary, Italy, Kosovo, Kuwait, Macedonia, Saudi Arabia, Somalia, Turkey, Uzbekistan, Zaire, and—most recently—Iraq. As Singer concludes, "It is not an exaggeration to note that wherever the US military goes, so goes Brown & Root."[59]

KBR did not originate as a PMC however, but rather as a Texas-based construction firm focused on the oil industry. In 1919, two brothers, George and Herman Brown, with financial backing from their brother-in-law, Dan Root, established Brown & Root. Initially conceived as a construction and engineering firm, the company prospered and grew. During World War II, Brown & Root had received numerous military contracts and had built hundreds of ships for the US Navy. Brown & Root personnel accompanied the US military in both Korea and Vietnam, building bases, roads, and harbors. However, during the 1970s Brown & Root exited the military business, focusing instead on energy projects.

In 1963 George Brown sold Brown & Root to Halliburton. This latter company, founded in 1919 by Erle P. Halliburton, began in Wilson, Oklahoma as the New Method Oil Well Cementing Company. Combined, the corporation experienced spectacular growth in the 1970s, associated with the expansion of oilfields under the North Sea and in the Middle East.[60] Later, however, during the recession of 1970s and a down-turn in oil market, the company formed Brown & Root Services subsidiary in 1986 as part of a diversification strategy. Following the Gulf War, Halliburton was contracted to bring burning oil wells under control, and Brown & Root selected to assess and repair the damaged public buildings in Kuwait.[61]

As part of its effort to diversity, in the early 1990s Brown & Root Services re-entered into military services.[62] In 1992 Brown & Root won a contract from the US Army's LOGCAP (Logistics Civil Augmentation Program) to work with the military in planning the logistical side of contingency operations; this was the first time the US military had ever contracted such global planning to a private organization.[63] In 1995 former US Secretary of Defense Dick Cheney joined Halliburton as President and Chief Executive Officer. Kellogg, Brown & Root is also known as Engineering and Construction Group, under which the military support operations of Brown & Root Services fall. Finally, with the addition of oil-pipe fabricator M.W. Kellogg (in 1998), Brown & Root became Kellogg, Brown & Root (KBR).[64]

Brown & Root's lucrative association with the US military has been challenged by another firm, the Reston, Virginia-based Dyncorp. Founded in 1946 by a group

---

59 Peter W. Singer, *Corporate Warriors: The Rise of the Privatized Military Industry* (Ithaca, NY: Cornell University Press, 2003), 136.

60 Singer, *Corporate Warriors*, 137.

61 Singer, *Corporate Warriors*, 138.

62 Anthony Bianco and Stephanie Anderson Forest, with Stan Crock and Thomas F. Armistead, "Outsourcing War: An Inside Look at Brown & Root, the Kingpin of America's New Military-Industrial Complex," *BusinessWeek Online*, September 15, 2003, http://businessweek.com/magazine/content/03_37/b3849012.htm (May 11, 2005).

63 Singer, *Corporate Warriors*, 138.

64 Bianco and Forest, "Outsourcing War."

of returning World War II pilots seeking to use their military contacts to make a living in the air cargo business, Dyncorp has grown to be the US's thirteenth largest military contractor by 2002. It was Dyncorp that worked under the Plan Colombia contract, wherein the company had 88 aircraft and 307 employees—139 of them American—flying missions to eradicate coca fields in Colombia.[65]

Custer Battles is a Fairfax, Virginia-based PMC named after and headed by Scott Custer and Mike Battles.[66] Custer, a former US Army officer, engaged throughout his career in clandestine operations in the Persian Gulf, Latin America, and Africa. He later attended, and received degrees in security studies, from Georgetown and Oxford Universities. Having worked as a defense consultant concentrating on counterterrorism and peace-operations initiatives, Custer founded a transnational corporation to provide security managers and operational staff across the globe in risk assessment, emergency planning, and crisis response. Custer also regularly appeared on Fox News as a military analyst.

Battles, likewise, had considerable military experience as well as working in the intelligence community. A West Point graduate with a Bachelor of Science in geopolitics, Battles served on special operations assignments throughout Eastern Europe and the Balkans. He also worked with the Central Intelligence Agency and, later, was a candidate for the US House of Representatives from Rhode Island.[67]

In the fall of 2002 the two men founded the security firm Custer Battles. Later, two other affiliates were established: Secure Global Distribution (SGD) and Med-East Leasing, both based in the Caymen Islands. Fred Rosen quotes the underlying motivations of Custer Battles from one of their in-house brochures:

> Iraq is a nation and marketplace wrought with challenges, obstacles, and malevolent actors. However, Iraq offers contractors, traders, entrepreneurs as well as multinational enterprises an unprecedented market opportunity. The ability to identify, quantify, and mitigate this myriad of risks allows successful organizations to transform risk into opportunity. Terrorist, sophistical criminal enterprises, political and tribal turmoil, and a lack of modern infrastructure present formidable challenges to companies operating in all areas of Iraq.[68]

SGD is engaged in logistical services, in effect a supply line to support security contractors. Various services provided by SGD include air cargo service, sea freight, trucking and distribution, equipment leasing, receiving services, warehouse services, and secure cargo escorts. Customers regularly include private contractors, government entities, and non-profit organizations.[69]

---

65 Chatterjee, *Iraq, Inc.*, 110-1.

66 The following is dervied from Fred Rosen, *Contract Warriors* (New York: Alpha, 2005), 144-48.

67 Rosen, *Contract Warriors*, 145.

68 Rosen, *Contract Warriors*, 145.

69 Rosen, *Contract Warriors*, 147-48.

Within months of its founding, Custer Battles was among the first contractors to enter Baghdad as part of America's reconstruction effort. By the end of June the contract awarded to Custer Battles to provide security for Baghdad's airport was the first time the company had provided security for an actual site. Two months later the company bid on and won a contract to support Iraq's currency distribution. The value of the contracts were worth approximately US$20 million.[70]

In many respects, Custer Battles has come to symbolize the exploitation of Iraq. In 2004 and 2005 reports surfaced indicating that the corporation not only gouged the CPA, but continued to win new government contracts after abuses were first alleged. Among the fraudulent activities include the billing of a US$33,000 food order at US$432,000 and the invoicing of an electric bill of US$74,000 at US$400,000.[71]

The use of PMCs is part of a more general trend of privatized warfare and labor mobility. Recent decades have witnessed a search by firms for greater flexibililty in how work is organized and labor is deployed. Such strategies include job rotation, flexible job arrangements, and the increased use of part-time (domestic or foreign) workers. International contract labor migration provides one alternative strategy for firms in their adjustment to cyclical and long-term structural changes. The use of contract workers also provides the necessary flexibility to address the spatial and temporary fluidity of military interventions.

David Phinney explains that "tens of thousands of contract workers have helped set new records for the largest civilian workforce ever hired in support of a US war." This work-force, according to Phinney, is pyramidal in organization, with the US government sitting atop. Just below the US administration is a layer of prime contractors like Halliburton, Dyncorp, Custer Battles, and Bechtel. These corporations typically out-source their labor needs, often to a layer of subcontracting companies—mostly based in the Middle East—including Prime Projects International, First Kuwaiti Trading and Contracting, Alargan Trading of Kuwait, Gulf Catering, and the Saudi Trading & Construction Company of Saudi Arabia. These subcontractors, in turn, hire workers from largely impoverished countries, including the Philippines, India, Pakistan, Sri Lanka, Nepal and Pakistan. Phinney identifies such a layered system not only cuts costs for the prime contractors, but also creates an untraceable trail of contracts that clouds the liability of companies and hinders comprehensive oversight by US contract auditors.[72]

As the occupation continues, stories of labor abuse continue to surface. One recent example is the construction of the US embassy in Baghdad. Likely to become the biggest, most fortified diplomatic compound in the world, the embassy is being built by the First Kuwaiti General Trading and Contracting (FKTC) construction firm.

---

70 Jason McLure, "How a Contractor Cashed in on Iraq," *Legal Times*, 4 March 2005, www.law.com/jsp/law/LawArticleFriendly.jsp?id=1109859526942 (27 February 2006).

71 McLure, "Contractor Cashed."

72 David Phinney, "Blood, Sweat & Tears: Asia's Poor Build US Bases in Iraq," *Corpwatch* 3 October 2005, www.globalpolicy.org/security/issues/iraq/reconstruct/2005/1003asiaspoor.htm (27 February 2006).

FKTC, a Kuwaiti-based company, was recently awarded a US$592 million contract to build the embassy, despite long-standing accusations of employee exploitation and coercion.[73]

The construction of the US embassy is carried out not by Kuwaiti or American workers, but rather contracted workers mostly from Nepal and the Philippines. During its construction, approximately 900 workers live and work at the job site under horrendous working conditions. As described by David Phinney, these workers live in cramped housing, eat poor food, and lack satisfactory medical care and safety gear. They normally work 12 hour shifts, seven days a week, driving trucks, cleaning latrines, collecting rubbish, running dining facilities and warehouses. Phinney concludes that "Without them, and the 'body shop' contractors that provide such laborers, the US and coalition military camps—virtually small cities—would shut down."[74]

Recently, allegations were charged against FKTC by Ramil Autencio and other Filipinos who were contracted by FKTC in late 2003 and early 2004. Originally recruited for employment by the Philippines-based MGM Worldwide Manpower agency, Autencio claims that he was hired to work at the Crown Plaza Hotel in Kuwait, only to be 'forcibly' pressured to work in Iraq. Numerous other Filipinos and Nepalis have accused the corporation of mistreatment, overwork, and underpayment.[75] Other examples include the charges against Custer Battles. As part of its contracts, Custer Battles has hired Filipino workers in support of its catering operations.

The occupation of Iraq thus brings to the fore the transnational movement of contract warriors for economic gain. This movement has been accompanied by another practice, namely the movement of contract workers. And whereas the United States was the major provider of military personnel and military-related corporations in the invasion and occupation of Iraq, the Philippines positioned itself as the major provider of contract labor.

Filipino Workers of the World

On 14 April 2003, as Pentagon officials declared a cessation of major combat operations in Iraq, Philippine President Gloria Macapagal-Arroyo signed Executive Orders 194 and 195. Through the first order, the president approved the formation of a public-private sector task force to coordinate the Philippines' participation in the post-war reconstruction of Iraq. EO-194, in particular, codified the established of the Public-Private Sector Task Force on the Reconstruction and Development of Iraq (hereafter referred to as the Task Force).[76] Operationally, the Task Force was

---

73 David Phinney, "Baghdad Embassy Bonanza," *CorpWatch*, 12 February 2006, www. globalpolicy.org/security/issues/iraq/contract/2006/0212embassy.htm (27 February 2006).

74 Phinney, "Baghdad Embassy."

75 Phinney, "Baghdad Embassy."

76 EO-195 created an additional task force to provide humanitarian assistance to Iraq. The initial size of this contingent was expected to be approximately 500, consisting mostly of

assigned the dual purpose of assisting "the participation of Philippine companies in the rehabilitation and development of the Iraqi infrastructure" and developing "procedures to expedite deployment of Philippine manpower and other services in the fulfillment of contracts." This conformed readily with the earlier neoliberal policies and transnational practices enacted by the Philippine Overseas Employment Administration (POEA) to increase the efficiency in the deployment of contract workers.[77]

That the Philippine government would rapidly attempt to deploy contract workers to war-torn Iraq should come as no surprise. Currently, the Philippines is the world's largest exporter of government-sponsored contract workers. Annually, approximately 800,000 Filipinos are deployed on temporary contracts—of six-month to two-year duration—to over 190 countries and territories. Proportionately, the largest number of Filipinos are deployed to the Middle East. In 2004, of the 933,588 Filipinos deployed, 352,314 (or 38 percent) were sent to the Gulf States. The Kingdom of Saudi Arabia remains the largest recipient, garnering a share of 20 percent of all deployed workers world-wide.

Overseas employment supports the Philippines' economy. The export of workers effectively matches the growth of new entrants to the Philippines' domestic labor market. As such, overseas employment has staved off, temporarily, higher levels of unemployment than would otherwise be expected. More significant is the volume of dollars that migrant workers inject into the country. Between 1975 and 1994, for example, Filipino overseas contract workers remitted approximately US$18 billion; this represented nearly 3 percent of the country's gross national product. Moreover, remittances have become a major component of the country's total export earnings. By 1992, for example, remittances as a proportion of overall export goods and services reached a record high of 12.7 percent. And finally, in aggregate terms, the amount of earnings contributed to the gross national product by remittances from 1975 to 1994 (the aforementioned US$18 billion) was approximately four times larger than the total foreign direct investment for the same period.[78] Most recently, in 2004 Filipino contract workers remitted a staggering US$8.5 billion.

The activities of the United States, its invasion and subsequent occupation of Iraq, illustrate well the business of war. Likewise the activities of transnational corporations such as Dyncorp and Halliburton reflect the business of war. But we must not discount the actions of other countries. If the US is the world's

---

military engineers, police, doctors, nurses, and social workers. This force, however, was pared down when the government learned that the United States would not foot the bill.

77 For a more complete discussion of the Philippines' participation in the Coalition of the Willing, see James A. Tyner, *Iraq, Terror, and the Philippines' Will to War* (Boulder, CO: Rowman & Littlefield, 2005); for a discussion on the Philippines' overseas employment program, see James A. Tyner, *Made in the Philippines: Gendered Discourses and the Making of Migrants* (London: Routledge, 2004) and Joaquin L. Gonzalez III, *Philippine Labour Migration: Critical Dimensions of Public Policy* (Singapore: Institute of Southeast Asian Studies, 1998).

78 Gonzalez, *Philippine Labour Migration*, 73-5.

supplier of warriors, the Philippines is the world's largest supplier of workers. It is no coincidence, for example, that all of the major corporations (i.e., Halliburton, KBR, Dyncorp) contracted to rebuild Iraq have lengthy histories of hiring Filipino workers. Thus, while acknowledgment must be made of the pervasiveness of military contractors in the occupation of Iraq, so too must acknowledgment be made of the thousands of contracted and sub-contracted laborers who work and (sometimes) die in the country.

The Philippines is the largest supplier of foreign contract workers for the US-led coalition. In mid-2005, for example, there were an estimated 6,000 Filipinos working in various coalition camps throughout Iraq. The dominance of Filipino workers in Occupied Iraq is not a coincidence. Indeed, the Philippines' participation in the Coalition of the Willing was fundamentally tied to a broader effort to export labor as well as America's overall approach to reconstruction.[79] Stemming from regulations established by the Bush administration, only United States companies were allow to bid on reconstruction contracts using US funds. In turn, only coalition countries were allowed to bid on contracts paid for by the Development Fund for Iraq.[80] And it was this aspect that established the Philippines' initial support for the invasion of Iraq.

Government involvement in the export of labor dates to 1915 when the incipient Philippine legislature passed Act 2486. This earliest act addressed issues such as child labor, recruiters' tax liabilities, and licensing obligations. However, early efforts were minimal, and the initiation of an effective government regulation of overseas employment did not materialize until the martial law regime (1972-1981) of President Ferdinand Marcos. Marcos's "New Society" was framed as a means to rectify existing social, political, and economic problems that were plaguing the country. Economic policies were realigned to attract new private investment as Philippine development policy became increasingly oriented toward export production. Marcos, in effect, ushered in the large-scale globalization of the Philippines' economy, albeit to serve his (and his cronies) personal aggrandizement. Following the success of other Asian states, such as South Korea and Singapore, Marcos promised his version of 'martial law, Philippine style.'[81]

Within a political environment of martial law, Marcos facilitated a major reorganization of labor relations within the country, changes that ultimately weakened the position of national labor. Marcos's New Society was predicated on an ideology that emphasized individual and national discipline and, concurrently, the sacrifice of personal liberties for economic development. From the outset, therefore,

---

79 Tyner, *Philippines' Will to War*.

80 Phillips, *Losing Iraq*, 139.

81 Philip F. Kelly, *Landscapes of Globalization: Human Geographies of Economic Change in the Philippines* (London: Routledge, 2000), 33; see also Walden Bello, D. Kinley, and E. Elinson, *Development Debacle: The World Bank in the Philippines* (San Francisco: Institute for Food and Development Policy, 1982); Gary Hawes, *The Philippine State and the Marcos Regime: The Politics of Export* (Ithaca, N.Y.: Cornell University Press, 1987).

Philippine overseas employment has been forwarded within neoliberal mind-set, as labor-related policies in the Philippines have sacrificed individual 'bodies' for a more 'global' vision of the nation-state.

The New Society was founded on a repressive labor policy that banned strikes, eliminated unions, and entailed a downward revision of existing labor protection standards.[82] The 1974 Labor Code, for example, as well as subsequent presidential decrees, favored foreign investors, selected members of the landed oligarchy, and Marcos personally. General Order 5, as a case in point, imposed a total ban on strikes and other forms of public assembly. This was later amended to limit the ban on strikes to only 'vital' industries, which included companies engaged in the production or processing of essential commodities or products for export. The Labor Code, likewise, permitted employers to pay new employees only 75 percent of the basic minimum wage during a six-month probationary period. By releasing workers after this period, multinational corporations effectively instituted a high turnover rate, thereby reducing labor costs even more. Marcos claimed that the loss of civil liberties was a regrettable but temporary price that Filipinos would have to pay for political stability, economic growth, and social reform. This price would increase over the years, in the form of abuse, exploitation, rape, and murder.

The export of labor readily conformed with the Marcos regime's policy of development diplomacy. This policy was predicated on the observation that less-developed countries contained large population bases as well as vital natural resources and that these could be used for development goals—hence, the use of the country's surplus labor and the high demand for labor, especially in oil-producing countries. Labor would become the Philippines' comparative advantage. Subsequently, the incipient Philippine 'migration industry' capitalized on newly emerging employment opportunities throughout the Middle East and Asia. Before the oil embargo of 1973-74, the Middle East labor market was characterized by a high degree of intra-regional, inter-Arab migration. Workers from oil-poor but relatively labor-rich states, such as Yemen, supplied labor to other states. Moreover, the scale of industrialization was such that regional migration could well meet any labor demands. However, increases in oil revenues following the embargo changed the Gulf labor market. Flush with petrodollars, many Middle East states, including Saudi Arabia, Kuwait, the United Arab Emirates, and Bahrain, initiated massive infrastructure projects.

With the massive infusion of capital into the region, relative and absolute labor shortages followed. Regional systems of migration could not meet the needs of capital. In response, Gulf states began recruiting foreign workers from East Asia, South Asia, and Southeast Asia. South Korea was an early supplier, followed by other countries such as Pakistan, Sri Lanka, Thailand, and the Philippines. Local governments were quick to realize the advantages of hiring non-Arab migrant

---

82 S. Kuruvilla, "Economic Development Strategies, Industrial Relations policies, and Workplace IR/HR Practices in Southeast Asia," in *The Comparative Political Economy of Industrial Relations*, ed. K. Wever and L. Turner (Madison: Industrial Relations Research Association Series, University of Wisconsin, 1995), 115-50.

workers. Apart from providing a source of skills in volumes no longer available in the Gulf, migrant workers from Asia were generally confined to isolated work camps and would then depart when the project was completed. This was especially attractive to the smaller, less populous states, such as the United Arab Emirates, that were concerned about a disproportionate number of foreigners living in their midst. Relatedly, there had been a growing uneasiness toward Arab migrants who would potentially decide to resettle after migration. This was not the case with Asian workers who, culturally estranged from the local populations, had no desire (or opportunity, in many cases) to stay and settle.[83]

The 1980s witnessed an increased neoliberal realignment of overseas employment. This was marked by the establishment of the Philippine Overseas Employment Administration (POEA). This state-run agency has emerged as the primary government apparatus tasked with regulating and monitoring of overseas employment. The overriding philosophy of the POEA was detailed in the inaugural issue of the *Overseas Info Series* (an in-house journal published by the marketing branch of the Preemployment Services Division of the POEA):

> A discriminate marketing approach guides the development of labor markets for Filipino overseas contract workers. Tapping non-traditional markets, whether skill-based or geographic-based, is geared towards high-benefit, high-growth areas suited for grooming a premium international image for the Filipino workers. Responding to future higher skills demand, it is the mission of our marketing program to equip itself with suitable surplus labor to fill demand trends. It is our further concern to ensure that no one market spoils our overall market image and/or block our entry to newly emerging opportunities. When net returns are ascertained to be unfavorable in any market segment, it becomes our task to recommend suspension and/or closure of such market and pour resources into better alternatives. It is the program's responsibility to protect the well-being of the Filipino workers.[84]

Although the final sentence makes passing reference to the 'well-being' of migrant workers, the overall tone is clear: market demand takes priority above all else. The Philippine government, through the policies and practices of the POEA, sought to fully incorporate itself into foreign labor markets. The presidential administrations of Corazon Aquino and Fidel Ramos continued these practices. Under the National Economic Development Authority's Medium-Term Development Plan for 1993-1998, for example, the Ramos administration's position was strong in stating that it would maximize the economic benefits, but it was weak on voicing its determination to minimize the social costs. Ramos did suggest that as more industries located in the Philippines—thereby generating more domestic-based employment opportunities—the export of labor may gradually diminish in importance.

---

83 Peter N. Woodward, *Oil and Labor in the Middle East: Saudi Arabia and the Oil Boom* (New York: Praeger, 1988), 9-10.

84 Philippine Overseas Employment Administration, "Market Development: Seeking Purpose and Promise for Filipino Skills," *Overseas Employment Info Series* 1 (1988): 5-9.

Substantial changes occurred in the 1990s following to highly publicized cases of migrant abuse. The first centered on Flor Contemplacion, a female domestic workers employed in Singapore. Contemplacion, in 1991, was charged with the double murder of another Filipina domestic worker and a four-year-old Singaporean boy. The Philippine media portrayed Contemplacion as a martyr, one of the many Filipinos and Filipinas who 'sacrificed' their lives for the greed of the Philippine government. During the weeks leading up to her scheduled execution in 1995 Ramos attempted to intervene, as did Philippine Catholic church leader Jamie Sin. These attempts were unsuccessful, however, and Contemplacion was executed on 17 March 1995. The second case focused on Sarah Balabagan, a fifteen-year-old girl, who was on trial in the United Arab Emirates for murder. Under-age, Balabagan had entered the country on a forged passport—she claimed to be twenty-eight years old in order to bypass POEA regulations. In July 1994 Balabagan was allegedly attacked and raped by her eighty-five-year-old male employer. At the conclusion of the trial, Balabagan was found guilty of manslaughter and sentenced to seven years' imprison. She was also, paradoxically, awarded a settlement of US$27,000 for being raped. Her case was appealed by both the prosecution and defense and in September 1995 the prison sentence was overruled. Instead, and surprisingly, Balabagan was sentenced to death. Following the outcry over the execution of Contemplacion, the Philippine government worked to secure Balabagan's release. Ultimately, the president of the United Arab Emirates intervened and the execution was dropped in lieu of a 'blood' payment of US$41,000 to be paid to the slain employer's family, one hundred cane lashes and a twelve-month prison term (of which she served approximately eight months).

These two events contributed to a significant reorganization, though not transformation, of the Philippines' overseas employment program. President Ramos, on 7 June 1995, signed Republic Act (RA) 8042, an act that emerged from the consolidation of House Bill 14314 and Senate Bill 2077. Hailed as the Magna Carta of overseas employment, the Migrant Workers and Overseas Filipino Act of 1995 signaled a discursive—but not practical—change in the government's approach to overseas employment. Specifically, RA 8042 was an avowal by the government that it did not promote overseas employment as part of the country's development program.

RA 8042 was a neoliberal watershed. Conforming with key tenets of neoliberalism, namely the promotion of an unrestricted free market, individual choice, the reduction of government regulation, and the promotion of private corporations, RA 8042 in theory would further entrench overseas employment in Philippine society while minimizing governmental obligations and culpability. As articulated in RA 8042:

> While recognizing the significant contributions of Filipino migrant workers to the national economy through their foreign exchange remittances, the State *does not promote overseas employment as a means to sustain economic growth and achieve national development.* The existence of the overseas employment program rests solely on the assurance that *the dignity and fundamental human rights and freedoms of the Filipino citizen shall not, at any time, be compromised or violated.* The State, therefore, shall continuously create

local employment opportunities and promote the equitable distribution of wealth and the benefits of development.[85] (Italics added.)

RA 8042 constitutes a neoliberal attack on the government regulation of overseas employment. Founded on the premise that people should be allowed maximum 'freedom' to pursue individual economic enterprises, RA 8042 advocates that 'free' labor markets should be the keystone of development strategies. Moreover, RA 8042 was a means to reduce government intervention and government culpability. Section 29 of RA 8042, for example, mandated the DOLE to formulate a five-year comprehensive deregulation plan on recruitment activities; section 30 mandated a gradual phase-out of all regulatory functions. As detailed by Rochelle Ball and Nicola Piper, the passage of the act was based on a strong endorsement of deregulating the responsibility of governmental institutions, supposedly in the name of worker welfare.[86]

In response to the passage of RA 8042 the POEA released a White Paper on overseas employment. Entitled *Managing International Labor Migration and the Framework for the Deregulation of the POEA*, this paper clarified the POEA's new role as 'manager' of migration. As specified in the paper, the POEA under deregulation was to assume a new philosophy of overseas employment. "Managing a global phenomenon," the White Paper asserts, "starts with understanding the philosophy of humankind, dynamics of migration, history and natural laws which cannot be repealed." The POEA's White Paper, in effect, went beyond a mere technical report on the operations of labor export to provide a holistic neoliberal justification and rational of overseas employment. Drawing on neoclassical economic ideas, infused with elements of social Darwinism, the white paper portrayed the Philippine state as being 'controlled' by the natural laws of globalization. The document continues that the "economic law of supply and demand is an irrepressible force in the global labor market," more so "now with the globalization era." Moreover, "this reality seems overshadowed by the application of national labour laws and administrative systems that perpetuate a pathological fallacy that labour migration is a program creation or innovation of government to address employment gaps."[87] In effect, the POEA has constructed overseas employment as a natural feature of globalization; consequently, people who seek overseas employment do so out of an individual cost-benefit decision-making process.

The White Paper positions the Philippines within the global economy as a natural supplier of labor. "Media sensationalism," the paper charges, "tends to overshadow

---

85 Philippine Overseas Employment Administration, *Migrant Workers and Overseas Filipinos Act of 1995: Republic Act 8042 and its Implementing Rules and Regulations* (Manila: Department of Labor and Employment, 1996), 2.

86 Rochelle Ball and Nicola Piper, "Globalisation and Regulation of Citizenship–Filipino Migrant Workers in Japan," *Political Geography* 21 (2002): 1013-34.

87 Richard R. Casco, *Managing International Labour Migration and the Framework for the Deregulation of the POEA* (Manila: Philippine Overseas Employment Administration, 1997), 2-3.

the fact that the global presence of Filipino labour is our strategic contribution to the global development, from which we reap net rewards central to the goals of our social economy. Managing means being able to objectively recognize and subsequently dominate our niche and comparative advantage ... "[88] The transnational practice of labor deployment is constituted as an inherent feature of the global economy. Just as other countries have been 'blessed' with abundant natural resources such as oil and gas, the Philippines has been bequeathed a surplus of labor. Structural conditions and practices, such as the promotion of export-oriented industrialization, unfair trade agreements, the expansion of agribusiness, and inequitable distributions of land-ownership, are ignored. And just as the United States must capitalize on its pre-ordained position as the world's hegemonic superpower, so too must the Philippines capitalize on its own market niche. This is to be accomplished not by the active promotion of migration but rather through the creation of a favorable, privatized, market environment. As expressed in the white paper, "As human rights with democratic ideology, our migrant workers assert their constitutional rights and exercise their faculty of judgement to satisfy their socioeconomic needs and wants. No amount of legislative or other structural barriers can effectively suppress these natural drives and human rights unless enforcement of such barriers is undertaken strictly under an authoritarian regime."[89]

The POEA has constructed migrant workers as empowered individuals who express their constitutional and human rights through spatial movement. This sense of empowerment is encapsulated in the concept of 'full disclosure'. Full disclosure "calls for the policy norm that: honesty is the best policy. To promote a culture of well-informed public is to stimulate a universal environment conducive to it." Full disclosure "is not a matter of recognizing and accepting one provision of the employment contract and reject the others. It is laying the cards on the table and when one makes a decision, he is primarily responsible for that decision."[90]

Through a neoliberal policy of full disclosure, all government and private institutions are tasked to provide all information—complete transparency—such as wages, working conditions, and so on, to empower potential workers to make informed decisions regarding overseas employment. The POEA seeks to empower individuals to make free decisions based on freely accessible an relative information. However, such information will still require verification; consequently, it is the responsibility of the POEA to provide minimal regulatory intervention to ensure the proper functioning of the global labor market. As such, the POEA maintains the prerogative to impose and enforce labor standards, but the final decision to that reached between the employer and employee, each with 'full' understanding of the costs and benefits. Logically, this policy considers migrant workers to be willing participants in overseas migration through free and rational choice; the accountability

---

88 Casco, *Managing International Labour Migration*, 4.

89 Casco, *Managing International Labour Migration*, 4.

90 Richard Casco, *Full Disclosure Policy: A Philosophical Orientation* (Manila: Philippine Overseas Employment Administration, 1995), 4.

of exploitation and abuse is correspondingly transferred to the migrants themselves. This constitutes a disavowal on the part of the government to assume the primary responsibility for migrant welfare. Such as neoliberal transformation undermines worker welfare through a state retreat from its responsibilities. The POEA, in effect, promotes itself as an organization committed to the protection of individual rights, liberties, and freedoms. This counters—or at least attempts to counter—the criticisms leveled against the government for sacrificing workers in the name of capital accumulation. Consequently, the state rhetorically upholds the democratic principles of free choice and the (nominally) freedom of movement.

The Macapagal-Arroyo administration has continued the neoliberal philosophies of full disclosure, transparency, and migrant empowerment. However, the Macapagal-Arroyo administration is also determined to more overtly utilize overseas employment as a means to sustain economic growth and to achieve national development. Eschewing earlier discourses of 'managed' migration, and of migration being an natural process, the current administration has whole-heartedly supported the intensification of labor export. Macapagal-Arroyo has consistently set as a target the number of one million workers to be deployed in a single year, and has promoted a philosophy based on increased efficiency and the removal of bureaucratic red tape.

Under the Macapagal-Arroyo administration, the POEA is to continue to serve as a regulatory agency, overseeing the day-to-day operations of the thousands of private recruitment agencies. The POEA, moreover, is to aggressively seek out new labor markets, and to diversity existing markets. Both government intervention and competition are to be removed from the marketplace as the Philippine state continues to maintain its hegemonic position as the world's leading exporter of labor.

The Philippines' participation in the Coalition of the Willing and the subsequent occupation of Iraq is a direct outgrowth of the country's neoliberal agenda of promoting transnational labor migration. It is to this effort that I now turn.

Reconstructing Iraq

The Philippines' current approach to overseas employment is characterized by a process I term *anticipatory reactivism*.[91] Policy-making in the Philippines has historically been reactionary when it comes to overseas employment. Maruja Asis attributes this to two factors: (1) governmental management is of recent vintage, with policies being fashioned without the benefit of historical precedents; and (2) overseas employment was envisioned to be stop-gap program. Consequently, the approach to policy-making was reactive: policies were introduced or modified as new challenges surfaced.[92]

---

91 Tyner, *Philippines' Will to War*, 46.

92 Maruja M.B. Asis, "The Overseas Employment Program Policy," in *Philippine Labour Migration: Impact and Policy*, edited by G. Battistella and A. Paganoni, (Quezon City, Philippines: Scalabrini Migration Center, 1992), 68-112; at 69.

With three decades of experience, the Philippines' migration industry has learned to *anticipate* future challenges and opportunities. Accordingly, the POEA attempts to develop contingency plans to be utilized in case a particular 'market' opens or closes for whatever reason. Labor market openings, for example, may be associated with major infrastructure projects initiated by economic restructuring—such as when Taiwan opened its economy to foreign labor. Openings may also arise as a result of military conflict, such as the war in Iraq.

For the United States, oil occupied center stage in the decision to invade Iraq. The Philippines' participation in the conflict illustrates another component. For the Philippines, as with other countries, the regulation of overseas employment was (and is) part of the business of war.

The signing of EO-194 and EO-195 are testimony to the anticipatory reactivism of the Philippine state. Constructed as military operations were still underway, certain policy-makers of the Philippine state worked fervently to capitalize on the soon-to-be destruction wrought on Iraq.[93] The newly-created Philippines' Task Force, for example, would function to market Philippine contractors, migrant workers, and service providers with foreign firms as subcontractors for projects related to the reconstruction of Iraq. In this manner, the Task Force would facilitate the arrangement of sub-contracts for Filipino construction workers in a process mirroring those established in the 1970s. Specifically, the Task Force was to provide representation on behalf of the Philippine government and the private sector with national, multinational agencies, and international private contractors involved in the reconstruction of Iraq; serve as a one-stop hub linking Philippine subcontractors with their primary contractors and major contractors; identify and qualify Philippine-based companies and other entities to ensure that service standards were competitive and of the highest quality; and develop procedures and review workers and labor service-oriented qualifications in Iraq with the POEA to expedite deployment.[94]

The Task Force exhibited considerable continuity and conformity with the basic philosophy toward overseas employment as articulated by the Macapagal-Arroyo administration, namely to maximize efficiency and rapidity in the deployment of workers. This translated into efforts to reduce government intervention and to promote Filipino labor as an attractive international investment. In its initial marketing efforts, the Task Force emphasized the Philippines' international standing in overseas employment—it serves as a model program for the International Labor Organization—and its large poor of skilled labor available for immediate

---

93 In *Iraq, Terror, and the Philippines' Will to War* I consider at length the morality underlying Macapagal-Arroyo's decision to support the Coalition of the Willing and the reconstruction efforts in Iraq. There I argue that Macapagal-Arroyo was not purely opportunistic in her decisions; indeed, I make the case that her policies reflect a more complex rationale based on a Catholic vision of peace and poverty alleviation.

94 This information was obtained via the Philippine Task Force for the Reconstruction of Iraq web-site, www.engagephilippines.com. The site no longer exists.

mobilization. Emphasis was also placed on the experience, expertise, and range of capabilities offered by Filipino labor.

EO-194 begins with the assertion that the Philippines, as a member of the United Nations, desires to contribute to both the immediate humanitarian needs of the Iraqi people and to the multinational effort for the long-term reconstruction of the country. The Task-Force, as principle group involved, was constructed as part of the larger global coalition that entered Iraq not for self-aggrandizement, but rather out of democratic and humanitarian ideals. Former Foreign Affairs Secretary Roberto Romulo, chair of the newly formed Task Force, explained, "The Philippines is extending humanitarian assistance to Iraq to the extent that our modern national budget will permit." He indicated that the Philippines was helping extensively in the work of reconstruction and rehabilitation through the Philippines' "greatest asset"— its "skilled labor force." Romulo also drew on a stock positioning of the Philippines as a humanitarian country, one that, although limited in its resources, was willing to provide its only viable resource: human labor. Romulo continued: "We are fortunate and we should all be thankful that the lives of our [overseas Filipino workers] in the Middle East are safe as a result of the decisive action in Iraq. Now we can turn quickly to doing what we do best: providing the best-skilled and most mobile labor force in the world to help rebuild Iraq."[95]

The Philippine state used its position as a member of the Coalition of the Willing to support its efforts to participate in the reconstruction of Iraq. Grounded in a sense of anticipatory reactivism, in March 2003 Ambassador Roy Cimatu, head of the Middle East Preparedness Team,[96] explained that he expected a total of fifty thousand to one hundred thousand jobs to result as a consequence of the war in Iraq; Filipino workers, given their history of employment in the Middle East—and especially from their long-association with US corporations such as Halliburton—were expected to be in high demand. Macapagal-Arroyo acknowledged that jobs for Filipinos were one of the benefits the country would reap from joining the US-led Coalition of the Willing.[97]

Beginning in early 2003, before the onset of military operations, the United States government identified fourteen American firms that were to be considered for contracts for the rebuilding of a postwar Iraq. Indeed, the Bush administration launched the bidding process for reconstruction in mid-February, a month before the invasion began and at a time when the United Nations was still trying to negotiate

---

95 Office of the President, "Task Force to Secure Organize RP Role in Iraq," April 20, 2003, www.news.ops.gov.ph/archives2003/apr20.htm (August 29, 2003).

96 The Philippines' Middle East Preparedness Team was created prior to the war in anticipation of the 1.4 million Filipino contract workers who would be displaced as a result of the conflict.

97 Ma. Theresa Torres, "100,000 Jobs Await Pinoys in Postwar Iraq," *Manila Times*, March 28, 2003, www.manilatimes.net/national/2003/mar/28/top_stories/20030328top2.html (April 9, 2004).

a compromise to avert war.[98] As reported in *The Guardian*, as the UN attempted to prevent war, the US government was actually *awarding* construction contracts to rebuild Iraq. Halliburton, for example, was awarded a contract to resurrect the Iraq oil-fields in the event of war-related destruction; other companies receiving no-bid contracts included Bechtel Corporation, Kellogg, Brown & Root, Stevedoring Services of America, and the Flour Corporation. Many of these corporations had long histories of hiring Filipino contract workers. Indeed, many of these companies participated in the rebuilding efforts following the 1991 Gulf War. Most recently, Halliburton had hired Filipino (and Asian Indian) workers to construct the detention facilities for 'terrorists' at Guantanamo Bay, Cuba.

Iraq was seen by many to be a potential gold rush, albeit gold per se was not involved. Rather, oil, labor and other related commodities were the prize. According to one industry executive, "It's a sensitive topic because we still haven't gone to war." However, the executive conceded, "These companies are really in a position to win something out of this geopolitical situation."[99] Indeed. The motivations for military intervention in Iraq were, even prior to the war, obvious to all parties involved: capital accumulation through the exploitation of a sovereign state. Admittedly, Iraq was controlled by a ruthless, murderous dictator. Still, participation in an international community—a global village—requires that certain legalities are followed. The United States blatantly eschewed international law in an attempt to capitalize on the authoritarian regime of Saddam Hussein. Other countries, including the Philippines, followed suit.

The Philippine state intensified its effort to secure employment contracts after the 'completion' of the war. In May 2003, for example, Macapagal-Arroyo used her state visit to the United States in part to help market Philippine labor. During her state visit, she met with top American contractors to ensure that the Philippines would share in the postwar reconstruction projects. Romulo said of the president's mission, "The task force needs to continue to aggressively market and prepare companies and workers for deployment."[100]

Romulo acknowledged that US$2.4 billion had already been allocated by the US Congress for the rebuilding efforts; some estimates placed the total cost as high as US$100 billion over a five-year span. Romulo also clarified that Philippine officials were hoping to capitalize on public workers and energy industry projects, as well as other non-core activities, such as information technology, finance and accounting, catering, and logistics." Romulo added that "it's not just a question of sending people, because there is scope to do a certain amount of back-processing here." In

---

98 "Iraq: Reconstruction and U.S. Interest," *Globe and Mail*, April 1, 2003, accessed through CorpWatch, www.corpwatch.org/news/PND.jsp?articleid=6228 (November 14, 2003).

99 Danny Penman, "USA: Firms Set for Postwar Contracts," *Guardian*, March 11, 2003, www.corpwatch.org/news/PND.jsp?articleid=5928 (November 14, 2003).

100 Gil Cabacungan, "President to Meet US Contractors during Visit," *Inquirer News Service*, May 1, 2003, www.inq7.net/brk/2003/may/01/text/brkofw_1-1-p.htm (November 14, 2003).

other words, Romulo intimated that employment opportunities in the Philippines may also be forthcoming in the postwar effort. As a case in point, he noted that Flor Daniel, one of the fourteen firms set to participate in the reconstruction of Iraq, already had employed about eight hundred Filipino engineers doing designs for overseas projects."[101]

Kuwait was targeted as a key entry point for overseas employment contracts. During May 2003 the Task Force dispatched a seven-person delegation to Kuwait, meeting with key officials of the Kuwaiti business chamber. Romulo indicated that "Kuwait has a long history of commercial and culture relations with Iraq and is poised to be the main jump-off point." He continued, "Clearly the Kuwaitis have a leg up on everybody else and are going in there; and in our various meetings they expressed their eagerness to do this in partnership with Philippine companies. They said they have the capital and the knowledge of the market and we have the skilled labor."[102] The Philippine state, moreover, shared a lengthy history of capital-labor exchange with Kuwait. In 2002 the Philippines' deployed more than 25,000 workers to Kuwait; remittances from these workers totaled in excess of US$27 million.[103]

Jun Campillo, spokesperson for the Task Force, indicated that the Philippines would offer services as subcontractors to British and US firms based in either Kuwait or Iraq. In this respect, the Philippines' participation in the Coalition of the Willing becomes even more substantive. As the Bush administration envisioned to privatize the war and subsequent reconstruction effort, US firms would need to sub-contract basic services, including catering, cooking, and construction. Preferential treatment for the awarding of sub-contracts would be based on previous experience as well as 'support' for the American-led mission. Campillo acknowledged, for example, that the Philippines, along with other countries of the Coalition of the Willing, were "positioning and trying to get contracts for their countrymen." He added that, during the 1980s, the Philippines had as many as thirty thousand workers in Iraq and that the Task Force hoped to exceed that figure.[104]

During the summer of 2003 the Task Force and the POEA worked to promote and facilitate its overseas employment program. The POEA, in particular, produced a list of qualified local service contractors and local recruitment agencies to serve as the manpower-sourcing reference for prime contractors. Also established were formal recruitment and hiring guidelines to facilitate deployment. In July the Task Force met with over one hundred public- and private-sector participants for a briefing on the Postwar Iraq Reconstruction Program. Speakers included Bruce Derrick,

---

101 Pia Lee-Brago, "RP to Gain from 14 US Firms Awarded Contracts to Rebuild Iraq," *Philippine Star*, May 5, 2003, www.philstar.com/philstar (November 14, 2003).

102 Mayen Jaymalin, "Kuwaitis Partnering with RP in Iraq Rebuilding Drive," *Philippine Star*, May 12, 2003, www.philstar.com/philstar (November 14, 2003).

103 For 2003 and 2004, according to Bangko Sentral ng Pilipinas, the central Bank of the Philippines, remittances received from Filipino workers in Kuwait totaled US$78 million and US$85 million, respectively; www.bsp.gov.ph/statistics/spei/tab11.htm (June 1, 2005).

104 "RP to Seek 30,000 Jobs in Iraq for Pinoys," *Philippine Star*, May 1, 2003, www. philstar.com/philstar/ (November 14, 2003).

country representative of KBR Halliburton-Philippines, and Wendy Chamberlain, ambassador and USAID assistant administrator for Asia and the Near East. Philippine representatives included administrators of the POEA, National Labor Relations Commission, and the Department of Foreign Affairs. At the meeting, Labor Undersecretary Manuel Imson presented an overview of the rules, programs, and services offered by the Department of Labor and Employment (DOLE) for the deployment of contract workers. He specified the following measures: First, the Task Force would provide a listing and profiling of licensed agencies and contractors; second, it would make available the quick processing of migrant workers to Iraq through the 'one-stop processing' center at the POEA; third, its labor attachés would conduct on-site project verifications; fourth, the DOLE's Technical Education and Skills Development Authority would train and certify Iraq-bound workers; and fifth, DOLE would provide orientation on assistance programs for workers. In conclusion, Imson urged participants to use the DOLE's Computerized National Manpower Registry System (CNMRS), a database of workers' skills and competencies; approximately 1.2 million Filipinos were in the restry. Imson added, "We have also opened in the CNMRS a special category for skilled Muslim workers in the southern Philippines."[105]

In October of 2003 representatives of the Philippines attended the International Donor's Conference for Iraq, held in Madrid, Spain. Over three hundred private-sector companies were in attendance, representing over seventy countries. Ali Allawi, interim trade minister of the US-installed Iraqi Governing Council (IGC), said, "The new Iraq will be above all a market-oriented economy. We expect and fully hope to achieve a stable, democratic, modern and progressive country in the very near future with the assistance of the international community and the assistance of the international private sector." He continued, "No sector of the economy will be closed to foreign investment with the exception of the oil sector."[106] The oil sector, of course, was reserved for the United States. However, the remainder of Iraq was considered open territory, laid bare to any and all proposals for business ventures. Thirty-eight countries pledged loans and grants totaling approximately US$35.9 billion (excluding humanitarian assistance, export credits, and guarantees). The remaining thirty-five countries were either 'scouting' for economic opportunities, lacked resources to make a presence, or were unable to pledge donations. The Philippines made no pledge of monetary assistance.

The Philippines was not alone in its attempt to capitalize on the reconstruction efforts. Indeed, Romulo warned, "We have to act quickly to establish our role because there are about 50 members of the coalition and most are angling for a piece

---

105 Department of Labor and Employment, "RP to Ensure Deployment of Skilled Workers to Iraq," July 16, 2003, www.dole.gov.ph/news/pressreleases2003/July/203.htm (November 14, 2003).

106 Associated Free Press, "Iraq Tells Investors It's Open for Business," *Manila Times*, October 24, 2003, www.manilatimes.net/national/2003/oct/24/business/20031024bus8.html (April 9, 2004).

of the action." He continued, "The best opportunity is for experienced Philippine companies and skilled workers because the primary contractors have first-hand experience with Philippine companies and skilled labor." Romulo also indicated that US firms would be outsourcing up to 70 percent of all labor needs, preferably to countries that were part of the Coalition of the Willing.[107]

Conclusion

Occupied Iraq constitutes a confluence of transnational material practices that facilitate the accumulation of capital. In this chapter I have positioned our understanding of Occupied Iraq though a focus on two dominant trends—the movement of contract warriors and contract workers into the country. Special attention has focused on the pre-war activities of the Philippines, as this country represents that world's largest provider of government-sponsored contract migrants. Moreover, Philippine labor has been intimately connected with the everyday machinations of American private military corporations and other transnational businesses.

The Philippines of course is not alone in its provision of contract labor. Many other countries, including Egypt, Nepal, and Sri Lanka have provided workers to participate in the reconstruction efforts of Occupied Iraq. Nevertheless, an emphasis on the Philippines is warranted because of its intricate organizational operations and pervasiveness of American military interventions.

In the following chapter I widen my scope of analysis to consider very broadly one particular consequence of these transnational flows of workers and warriors, namely the abduction of contract workers by Iraqi insurgents. Chapter 4, consequently, serves as the apex of *The Business of War*. In particular, I consider how the various empirical processes that contributed to the occupation of Iraq—America's continual territorial expansion, the Philippines' quest for ever-wider labor markets—are manifest in the violence enacted toward contract workers. Simply put, I maintain that each abducted worker may be viewed as the personification of a militant neoliberalism.

---

107 Lee-Brago, "RP to Gain."

Chapter 4

Spaces of Political Subjugation

*"And sometimes what he has to do for the state is to live, to work, to
produce, to consume; and sometimes what he has to do is to die."*
Michel Foucault[1]

In the mid 1990s the Ejército Zapatista de la Liberación Nacional (EZLN, or
Zapatistas Army of National Liberation) came into prominence. The movement
began on 1 January 1994, a date set to coincide with the establishment of the North
American Free Trade Agreement (NAFTA). Separatist in orientation, the broader
goals of the Zapatista have included agrarian and social reform. Popular with the
poor and indigenous peoples of southern Mexico, the movement has attracted
a global following. This has been accomplished, in part, through the use of the
Internet, cyber-technologies, and other and practices to promote their ideas.

Relatively small and peaceful, the Zapatista movement has served both as a
catalyst and a clarion call for a broader anti-globalization movement The leader of the
Zapatistas, Subcomandante Marcos, became a media cause celebre, interviewed by a
variety of media like *Vanity Fair*, *60 Minutes*, and *Time* magazine. Oliver Froehling
writes that "With the Chiapas revolt, a minor province in Mexico made headlines
and refocused world attention for a few days on the problems of indigenous people,
taking the spotlight at a time of celebrated globalization."[2]

The Chiapas Revolt and the emergence of the EZLN finds semblance with other
anti-neoliberal movements. In 1999, for example, as officials from over 130 countries
assembled in Seattle, Washington for meetings of the World Trade Organization
(WTO), thousands of protestors themselves assembled on the city streets. The
'Battle of Seattle', as the event became known, focused the world's attention on the
social dimensions of economic globalization and issues of social justice and equality.
According to the protestors, the WTO—as the primary organization tasked with
promoting free trade—signified all that was wrong with neoliberal globalization.[3]

---

1   Michel Foucault, "Technologies of the Self," in *Ethics: Subjectivity and Truth,
Essential Works of Foucault, 1954-1984*, Volume 1, edited by P. Rabinow (New York: The
New Press, 2000), 409.

2   Oliver Froehling, "The Cyberspace 'War of Ink and Internet' in Chiapas, Mexico,"
The Geographical Review 87(2): 291-307; at 291.

3   Bruce D'Arcus, "Globalization and Protest: Seattle and Beyond," *WorldMinds:
Geographical Perspectives on 100 Problems*, edited by Donald G. Janelle, Barney Warf, and
Kathy Hansen (Boston: Kluwer, 2004), 21-24.

These anti-globalization movements, and many others, would manifest in the world-wide demonstrations denouncing the American-led invasion and occupation of Iraq. As Iain Boal and his co-authors write, the marches swept westward with the sun: from Melbounre and Sydney to Rome, Tokyo, London, Paris, Madrid, New York and San Francisco. Around the world an estimated fifteen to twenty million people made known their opposition to the military neoliberalism of the United States and its Coalition members. Boal et al. write that "It was a world-historical moment. Never before had such masses of people assembled, against the wishes of parties and states, *to attempt to stop a war before it began.*"[4]

The invasion and occupation of Iraq is but the latest manifestation of a military-supported neoliberal expansion. And it was this aspect that provided the foundation for much of the anti-war protests around the world. With chants of 'no blood for oil', critics of the war voiced their moral opposition to march toward war. It was argued by the protestors that the transnational practices enacted by the United States in Iraq were (and remain) colonial in intent and consequence. And it is for this reason that Derek Gregory introduces the concept of the 'colonial present'.[5] Gregory explains that he speaks about the colonial present because he wants to "retain the active sense of the verb 'to colonize': the constellations of power, knowledge, and geography" that "continue to colonize lives all over the world."[6] For Gregory, "the triumphal show of colonialism ... and its effortless, ethnocentric assumption of Might and Right are visibly and aggressively abroad in our own present. For what else is the war on terror other than the violent return of the colonial past, with its split geographies of 'us' and 'them,' 'civilization' and 'barbarism,' and 'Good' and 'Evil'?"[7]

Geography has always been a part of the struggle for political, economic, and social rights. Indeed, as Ruth Wilson Gilmore articulates, a geographic imperative lies at the heart of every struggle for social justice. That these struggles are manifest on the landscape should come as no surprise. Occupied Iraq became a colonial space; a de-humanized site of domination and resistance. The pictures have become disturbingly too familiar: charred bodies, bloated, bloodied corpses. The remnants of humanity littered across the battlefield that once was Baghdad. Or Fallujah. Basra. Nasariyah. The twisted metal of tanks and cars, tossed among the rubble of former homes and hospitals, schools and markets. Such is the colonial present inscribed on the landscape of Iraq.

But if we agree that the occupation of Iraq is part and parcel of same process, namely a globalized capitalist expansion, how then do we respond to the insurgency in Iraq? Are not the insurgents actively fighting against the same neoliberal agenda that protestors in New York, San Francisco, and London are condemning? In Iraq,

---

4   Iain Boal, T.J. Clark, Joseph Matthews, and Michael Watts, *Afflicted Powers: Capital and Spectacle in a New Age of War* (London: Verso, 2005), 1-3; italics in original.

5   Derek Gregory, *The Colonial Present: Afghanistan, Palestine, Iraq* (Malden, MA: Blackwell, 2004).

6   Gregory, *Colonial Present*, xv.

7   Gregory, *Colonial Present*, 10-11.

insurgents demand the removal of an American military presence. They express the desire for self-determination, and for the sovereignty of Iraq to be fully restored. Their actions, in short, constitute an anti-colonial resistance movement.

The study of resistance has, in recent years, captured the attention of Geographers and other social scientists. Indeed, as Tim Cresswell remarks, "resistance may well be the central theme of contemporary social and cultural geography." Therein lies the problem. Suddenly, every behavior may be viewed as an act of resistance. Cresswell laments that if "an act such as an armed insurrection or a general strike is equated with the act of farting in public or telling jokes about the boss," then "resistance is in danger of becoming a meaningless and theoretically unhelpful term." A second danger is that, too often, "resistance seemed irrevocably heroic and infused with positive moral value."[8] There is a tendency, in effect, to romanticize the resisters. But what about the suicide bombers in Iraq and elsewhere? What of the abductors of hostages in Iraq and elsewhere? Insurgents have utilized kidnappings and car bombs, assassinations and suicide attacks, rocket propelled grenade attacks and road-side bombings. The acts committed by the insurgents are clearly not synonymous with ironic forms of dress, hairstyles, or culinary tastes. From the perspective of most Americans, these actions do not reflect a sense of 'positive moral values.'

I am in agreement with Cresswell that we need to view the power of resistance not as a potent symbol of subaltern freedom, but as an indicator and diagnostic of power. Power, from a Foucauldian stand-point, is not something that is possessed by a dominant political or military force. Power, instead, is exercised; it is relational and, consequently, circulates. Therefore, as Cresswell explains, resistance thus reconceptualized is not romaticized as an indicator of power's absence—a zero-sum game between the weak and the strong, the powerful and the powerless—but rather as evidence for power's existence and an intervention that serves to delineate the mode of power in question.[9]

Joseph Nevins, in his passionate and incisive study of the mass violence that ripped apart East Timor between 1974 and 2002, asks, "Why do we pay greater heed to some atrocities and not others?" In answering his own question, Nevins argues that geographic proximity, power, and social distance must be considered. Our understandings of events is intimately associated with the representation of those events. Furthermore, social distance and geographic distance combine to make the plight of others more peripheral and, by extension, less relevant to our own lives. Ironically, this comes at a time when geographic distance—through such technological innovations as the Internet—has been reduced. Nevins notes that "we are able to operate over great distances but often with little actual interface with those affected by what we do." And for Nevins, this 'paradox of modernity' comes with a high price, for it is through this process that we evaluate the salience of atrocities to

---

8   Tim Cresswell, "Falling down: resistance as diagnostic," in *Entanglements of Power: Geographies of Domination/Resistance*, edited by Joanne P. Sharp, Paul Routledge, Chris Philo, and Ronan Paddison (London: Routledge, 2000), 256-68; at 258-9.

9   Cresswell, "Falling down," 264.

our own lives. Nevins writes that "modernity brings people around the world closer together while at the same time increasing our ability to deny our responsibility for the effects we have on others—especially those geographically distant." Consequently, as Nevins concludes, such 'erasures of complicity' "facilitates impunity for those culpable of bringing about gross levels of human suffering while undermining efforts to ensure restitution for the victimized population."[10]

Occupied Iraq is a space of struggle; it is a site of domination, resistance, and violence. It is also about the framing of domination and resistance. In his magisterial *Discourse on Colonialism*, Aimé Césaire charged that "no one colonizes innocently ... no one colonizes with impunity."[11] Fanon concurred. For him, the first encounter of colonialism was marked by violence.[12] In Iraq, this was a spectacular display of 'shock-and-awe', a glib phrase that attempts—but cannot gloss over—the widespread death and destruction of lives. Derek Gregory explains, however, that American and British audiences were not to witness these terrifying displays of arrogant violence. He writes that "In order to advance from the grounds for killing into the killing grounds themselves, imaginative geographies were mobilized to stage the war within a space of constructed visibility where military violence became—for these audiences at least—cinematic performance."[13]

We need not agree with the tactics and violence of the insurgents to understand their concerns. Indeed, according to Cresswell, "such an analysis of power and resistance does not need to make any moral claims about the identity of the resister and the oppressor. A model of resistance as a diagnostic of power makes no investment whatsoever in the subject position of the agents—it simply uses their acts as evidence for various modes of power, including the power of resistance itself."[14] And it is from this perspective that I approach the political subjugation of hostages in Occupied Iraq.

Placing Occupied Iraq

On April 18, 2005 Paul Bremer, former head of the Coalition Provisional Authority (CPA) and tasked with the reconstruction of Iraq, addressed an audience of students and faculty at Clark University. Bremer began with a joke about three hostages held by terrorists.[15] Such a callous and seemingly non-chalant attitude toward the plight

---

10   Joseph Nevins, *A Not-So-Distant Horror: Mass Violence in East Timor* (Ithaca, NY: Cornell University Press, 2005), 11-13.

11   Aimé Césaire, *Discourse on Colonialism*, translated by Joan Pinkham (New York: Monthly Review Press, 2000 [1955], 39.

12   Frantz Fanon, *The Wretched of the Earth* (New York: Grove Press, 1963), 36.

13   Gregory, *Colonial Present*, 198.

14   Cresswell, "Falling down," 266.

15   "Bremer Speaks at Clark, 100 Protest," April 19, 2005, http://worcester.indymedia. org/print.php?id=1086 (September 14, 2005).

of other human beings speaks volumes about the place of the Iraqi occupation in American foreign policy.

Diamond, among others is highly critical of Bremer. For Diamond, "Bremer spurned the appeals of a wide range of Iraqis—including many who were cooperating with us—and of the US mission to transfer authority quickly to an Iraqi interim government, and he proceeded to reshape Iraq through an occupation that he led, and over which he exerted tight, indeed, almost total, control." Moreover, it was Bremer and other top American officials who "proved unwilling to incorporate the players—Ba'athists and Arab nationalists—who would have been capable of defusing the Sunni-based resistance, and who were, in fact, sending signals that they wanted to talk directly to the United States."[16]

Similar to the Bush administration, Bremer thought that the United States confronted, in Iraq, a struggle between good and evil, and he therefore resisted the idea of negotiating with representatives of the insurgent forces. His view was that the insurgents were 'evil-doers' who should be captured and killed—as if there were only a finite number of them and that they were not being driven by a calculous of rational interest.[17] These Iraqi participants were ignored, essentially, because they would have understandably balked at the neoliberal reforms proposed by the CPA. Furthermore, these participants understood also that neither democracy for Iraq nor liberation of the Iraqi people was not the goal. Rather, Coalition forces sought a more basic territorial control over a strategic location in the Middle East and, by extension, a global hegemony. It is hardly surprising indeed that a survey of Iraqis conducted in September 2003 found that only five percent thought that the United States invaded to assist the Iraqi people; only one percent believed that the invasion was to establish democracy. Nearly half of those polled, however, believed that the United States occupied Iraqi to steal Iraq's oil.[18]

Timothy Oakes conceives of 'place' as consisting of two components: as a site of both meaningful identity and immediate agency. He elaborates that place, as a site of meaningful action for individuals, is action derived from linkages across space and time which makes place more of a dynamic web than a specific site or location.[19] Occupied Iraq, accordingly, is not a joke (except maybe to Bremer) but instead a place of transnational practices and neo-liberal discourses; it is a web of capitalist weaving, derived from the particular strategies and agency of foreign governments, multinational corporations, 'terrorist' networks, and private individuals. It is also the place of interactions: between contract workers, coalition soldiers, and Iraqi insurgents.

---

16 Larry Diamond, *Squandered Victory: The American Occupation and the Bungled Effort to Bring Democracy to Iraq* (New York: Henry Holt and Company, 2005), 294-295.

17 Diamond, *Squandered Victory*, 300.

18 Diamond, *Squandered Victory*, 25.

19 Timothy Oakes, "Place and the paradox of modernity, *Annals of the Association of American Geographers*, 87(1997): 509-31; at 510.

Governments speak of the 'good' that accompanies colonialism. And Coalition members today routinely speak of the milestones reached: elections, security force training, rebuilt infrastructure. But the silences are as significant as the rhetoric. Césaire succinctly captured the hypocrisy of colonialism: "They talk to me about progress, about 'achievements,' diseases cured, improved standards of living; they throw facts at my head, statistics, mileages of road, canals, and railroad tracks; they dazzle me with the tonnage of cotton or cocoa that has been exported, the acreage that has been planted with olive trees or grapevines."[20] But where occupying forces see progress, Césaire sees, instead, "societies drained of their essence, cultures trampled underfoot, institutions undermined, lands confiscated, religions smashed, magnificent artistic creations destroyed; food crops destroyed, malnutrition permanently introduced, agricultural development oriented solely toward the benefit of the metropolitan countries; the looting of products, the looting of raw materials."[21] One immediately thinks of the 100 directives of Bremer, the reorientation of Iraq's economy as a dependency of the United States, the forcible opening of the Iraqi economy to western intervention.

Occupied Iraq is a colonial landscape, one that would be immediately familiar to Fanon or Césaire. Indeed, if one wanted to understand—to witness, to experience, to re-live—the horrors and barbarism of past colonialisms, one need only travel to Iraq. The graphic display of mutilated bodies of Iraqi women, men, and children—displayed and consumed on CNN, FOX News, and other media outlets—evoke nightmarish images of the rubber terror and the dismemberment of African bodies in the Belgian Congo, the massacres of East Timor, or any other instance of colonialism.

The horrific ugliness of colonialism is present in Occupied Iraq. It is revealed in the images of dead bodies, of children blown up while attending school, of mothers gunned down while buying vegetables at street stalls, of contract workers shot while repairing wells for drinking water. It is revealed in statistics, grim numbers that persistently increase day by day. It is estimated that the devastation of the invasion and occupation has resulted in hundreds of thousands Iraqi deaths. During the first eighteen months of the occupation, for example, surveys conducted by researchers at Johns Hopkins, as well as the Red Cross, indicated that approximately 100,000 Iraqi civilians had died. The majority of deaths resulted from the numerous Coalition-led air strikes on towns and cities.[22] The seige of Fallujah alone resulted, conservatively, in over 800 Iraqi civilian deaths. It is revealed mostly though in the misery and suffering of a dehumanizing process that is too often explained away by vague statements of liberty, freedom, and democracy.

Occupied Iraq is a Manichean world, a place cut in two: it is a place of colonial settlers (i.e., the Coalition Provisional Authority, oil companies, and so forth) and colonized Others (i.e., the ubiquitous and faceless Iraqis variously conceived as

---

20 Césaire, *Discourse on Colonialism*, 42-3.

21 Césaire, *Discourse on Colonialism*, 43.

22 Phyllis Bennis, "On the eve of the elections," *Institute for Policy Studies*, October 31, 2004, www.ips-dc.org/comment/BEnnis/tp24eleceve.htm (September 29, 2005).

insurgents or as collateral damage). And yet the framing of the conflict is decidedly unequal—from both the perspective of the insurgency and well as the Coalition forces. Neither is free from their culpability of Occupied Iraq. As Césaire, Fanon, and even George Orwell understood, colonization dehumanizes both the colonizer and the colonized. At the level of the human body, there is no difference between the massive bombings that comprised the Coalition's 'shock-and-awe' campaign, the tortures at Abu Graib, or the beheadings of hostages. Césaire is succinct on this matter: "wherever there are colonizers and colonized face to face" there is "brutality, cruelty, sadism, conflict."

The struggle for Iraq is a colonial/anti-colonial struggle and, from this perspective, the conflict is immediately local. Glenn Perusek explains that "In contrast to the steady diet of demonizing rhetoric about the insurgent opposition served up to American soldiers and the American public alike, the foreign policymakers and intelligence community recognize full well that the opposition is broad-based" and is composed of students, former soldiers, intellectuals, tribal youths, farmers and so on. Moreover, the fundamental objective of the opposition (in a collective sense) is the immediate withdrawal of Coalition forces and national independence.[23]

It is important to acknowledge therefore, as Perusek and others argue, the insurgency in Iraq is neither monolithic nor united. It is not controlled by a single individual nor it is guided by a singular ideological vision.[24] David Phillips concludes, however, that "US officials did not come to grips with the nature of the opposition. They refused to acknowledge the broad base of support for the insurgency among Iraqis."[25] American officials, in fact, failed to admit that the insurgency in Iraq was not the workings of a small group of 'extremists' or 'foreign terrorists' but instead an anti-colonial movement.

Despite the proclamations by most members of the Bush administration, the anti-colonial insurgency is composed of various factions, some vying for their own particular positions in a post-Coalition Iraq, others seeking to strike a blow at Western rule in general. Two Baghdad-based journalists, Samir Haddad and Mazin Ghazi, describe the opposition as "a mixture of Islamic and pan-Arab ideas that agree on the need to put an end to the US presence in Iraq."[26] Haddad and Ghazi, furthermore, divide the resistance into three main divisions: Sunni resistance groups, Shia resistance groups, and Ba'athist factions. The primary Sunni resistance groups include the 'Iraqi National Islamic Resistance,' the 'National Front for the Liberation of Iraq,' and the 'Iraqi Resistance Islamic Front.' The objectives of the Iraqi National Islamic Resistance, formed on July 16, 2003, is to liberate Iraqi territory from foreign

---

23 Glenn Perusek, "The U.S. occupation and resistance in Iraq," *New Politics*, www. wpunj.edu/~newpol/issue38/perusek38.htm (June 9, 2005).

24 Perusek, "U.S. occupation."

25 David Phillips, *Losing Iraq: Inside the Postwar Reconstruction Fiasco* (New York: Westview Press, 2005), 201.

26 Samir Haddad and Mazin Ghazi, "An inventory of Iraqi resistance groups," *Al Zawra* (Baghdad), September 19, 2004, accessed from *Socialist Unity*, www.socialistunitynetwork. co.uk/voices/hostages.htm (June 9, 2005).

military and political occupation and to establish a liberated and independent Iraqi state on Islamic principals. This group has been responsible for attacks on Coalition forces mostly in the area west fo Baghdad, in the regions of Abu-Ghurayb, Khan Dari, and Fallujah. It claims to have carried out an average of 10 operations per day between July 27 and August 7, 2004. The most prominent of attacks were the shooting down of a helicopter in the Abu-Ghurayb region on August 1, 2004 and the shooting down of another helicopter near Fallujah on August 9, 2004.[27]

The National Front for the Liberation of Iraq was formed earlier, just days after the occupation of Iraq in April 2003. Composed of nationalists and Islamists, its activities are concentrated in Arbil and Karkuk in northern Iraq, in Fallujah, Samarra, and Tikrit in central Iraq, and in Basra and Babil in the south. The Iraqi Resistance Islamic Front, the newest of the Sunni resistance groups, announced its existence on May 30, 2004. This group claims to have carried out dozens of operations against US occupation forces, including the shelling of the occupation command headquarters and the semi-daily shelling of the Mosul airport. Additional Sunni resistance groups include the Hamzah Faction (appearing October 10, 2003); the Iraqi Liberation Army (July 15, 2003); the Awakening and Holy War; the White Banners; and the Al-Haqq Army.[28]

The main Shiite resistance group is the Al-Mahdi Army of Shia cleric Muqtada al-Sadr, known also as the al-Sadr group. Formed in July 2003 by Shiite leader Muqtada al-Sadr, the group quickly gathered between 10,000 and 15,000 youths, the majority of whom were from the poor of al-Sadr City in Baghdad. Sadr, a leading figure of the resistance, was strongly opposed the occupation. The al-Sadr group, however, did not initially engage in resistance fighting. Only in the spring of 2004 did he begin to speak of armed resistance. The greatest confrontation between his militia and occupation forces occurred in August 2004 in al-Najaf. Fighting continued for three weeks until a cease-fire agreement was signed.[29] While opposed to the occupation, Sadr has denounced car bombings and assassinations and the taking of journalists as hostages. He has also criticized as impious the beheading of foreigners.[30] A second Shiite group, the Iman Ali Bin-Abi-Talib Jihadi Brigades, appeared on October 12, 2003. It has vowed to kill any soldiers associated with the occupation and has threatened to expand the battleground to other countries if they were to send troops.[31]

Ba'athist factions include those groups that are loyal to the Ba'ath Party and the previous regime of Saddam Hussein. These groups, including al-Awdah (The Return) and Saddam's Fedayeen, are seen as largely responsible for financing resistance operations, rather than engaging in actual combat operations. Reports have surfaced

---

27 Haddad and Ghazi, "Iraqi resistance groups."
28 Haddad and Ghazi, "Iraqi resistance groups."
29 Haddad and Ghazi, "Iraqi resistance groups."
30 Perusek, "U.S. occupation."
31 Haddad and Ghazi, "Iraqi resistance groups."

that many of these members have, however, shifted loyalties to other Islamic and national resistance groups.[32]

Lastly, Haddad and Ghazi identify nine smaller groups that principally engage in the abduction and killing of hostages. Smaller and less organized the most of the aforementioned groups, these factions are relatively localized. Groups include the Assadullah Brigades, the Islamic Retaliation Movement, the Islamic Anger Brigades, the Al-Tawhid wa al-Jihad Group and its military wing known as the Khaled-Bin-al-Walid Brigades, the Black Banners Group, the Abu-Mus'ab al-Zaraqawi Group, the Islamic Army in Iraq, and the Ansar al-Sunnah Movement. The last four groups are believed to be related in ideology, if not in practice, to Al Qaeda.[33]

Compared to twentieth century insurgent oppositions to occupation, the Iraqi opposition has developed quickly, and with minimal international support.[34] Regardless of the statements emanating from the White House, the insurgency in Iraq is not dominated by transnational networks of 'terrorist' organizations. Recent studies indicate that foreign fighters probably constitute only about 4 to 10 percent of the estimated 30,000 insurgents.[35] Indeed, the insurgency has, in some respects, benefitted more from the oversights of the Coalition government than outside support. Consider, for example, the staggering amounts of weapons and ordinance that have been acquired by the insurgents from weapons caches within Iraq. After the US-led invasion, more than 760,000 pounds (380 tons) of lethal explosives, most of which can be used to make improvised explosive devices (IEDs), were stolen from the al-Qaqaa weapons depot in Iraq. The site, known to US authorities to be the main high explosives storage facility in Iraq, was left unguarded by coalition forces.[36]

The atrocities associated with the insurgency, furthermore, have steadily increased with the passage of each 'milestone' commemorating Iraq's supposed liberation. These include, for example, "the capture of Saddam Hussein in December 2003; the transfer of sovereignty in June 2004; and the holding of elections in January 2005. By these measures, according to the Bush administration, the war in Iraq— notwithstanding the failure to find weapons of mass destruction, the debacle of US military personnel's abuse of prisoners at Abu Ghraib, and the lack of security for civilians—has driven a successful effort to promote liberty and freedom abroad."[37]

According to Glenn Kutler, fatality figures tell a different story. During the period of the initial occupancy (c. April 10 - September 11, 2003), there were 177 fatalities to Coalition forces, averaging eight per week. These figures include the 23 August

---

32 Haddad and Ghazi, "Iraqi resistance groups."

33 Haddad and Ghazi, "Iraqi resistance groups."

34 Perusek, "U.S. occupation."

35 Tom Regan, "The 'myth' of Iraq's foreign fighters," *The Christian Science Monitor*, September 23, 2005, www.csmonitor.com/2005/0923/dailyUpdate.html (September 27, 2005).

36 Bennis, "Eve of elections."

37 Glenn Kutler, "U.S. Military Fatalities in Iraq: A Two-Year Retrospective," *Orbis* 49 (2005): 529-544; at 530.

explosion of a truck bomb outside the United Nations' headquarters in Iraq, killing Sergio Vieira de Mello and twenty-one others.[38]

Internationally, a significant signpost was reached on 22 May 2003 when the United Nations Security Council recognized the US and the UK, under international law, as 'occupying powers' in Iraq. Resolution 1483 called on the US and the UK to promote the welfare of the Iraqi people through an effective administration of the territory, and to create conditions for Iraqis to freely determine their future. The Bush administration viewed this as a positive step forward, but the Iraqi populace viewed it as the formal establishment of foreign occupation.[39] At this point, the simmering insurgency became a de-colonization movement. Indeed, Kutler identifies that at this point the organized insurgency came into full view. During the next ten-week period Iraq experienced 136 fatalities, with a weekly average number of deaths reaching 14.[40]

In response to mounting resistance, between 21 November 2003 and 6 March 2004 US and Coalition forces unleashed massive counter-offensive campaigns. This period also saw a transformation of the insurgency, as non-Americans—and especially Iraqis—were increasingly targeted. During this period 362 Shiites, Kurds, and Iraqi police offices, along with 7 Spanish and 2 Japanese civilians were killed. Bremer at this point viewed the increased attacks as symptomatic of a dying insurgency. Kutler disagrees, suggesting instead that that "the targeting of Iraqis signaled the beginning of a calculated effort by the insurgents to expand the scope of their attacks by engaging in sectarian violence and targeting collaborators and others in lieu of Americans." He maintains that "At a time when American fatalities had declined—perhaps because US forces were confined to base or otherwise too well defended—deaths of non-Sunni Iraqis and others skyrocketed."[41]

Despite the increased deaths associated with the occupation, the Bush administration has consistently downplayed, if not out-right ignored, the mounting casualties. Media images of dead soldiers, body bags, and flag-draped caskets have been minimized and the president (through the summer of 2005) attended no military funerals. The president's minimalist comments about fatalities has been standard administration practice.[42] Kutler elaborates that "the president's practice of avoiding mention of fatalities is politically pragmatic. Indeed, it has been part of a broad strategy to distance the electorate from the concrete realties of the war." He continues, noting that "The effort to distance the public physically from the war has been paralleled by an effort to create moral distance from the war." To this end, despite "the disturbing implications of Abu Ghraib, non-existent WMD, and Iraqi civilian casualties, the administration positioned these and similar issues as

---

38 Kutler, "U.S. Military Fatalities," 533.
39 Diamond, *Squandered Victory*, 38.
40 Kutler, "U.S. Military Fatalities," 534.
41 Kutler, "U.S. Military Fatalities," 534.
42 Kutler, "U.S. Military Fatalities," 539.

abstract and remote from the public, enabling voters to rationalize a vote for the status quo."[43]

Iraqis have not been so distanced from the occupation. As such, the increased abductions during the spring and summer of 2004 were an indication of underlying resentment toward American occupation.

Spaces of Struggle

Located to the west of Baghdad, in the province of Al-Anbar, Fallujah is a medium-sized city of approximately 300,000. In the early stages of the occupation the US Army's First Infantry Division converted one of Fallujah's schoolhouses into a barracks. Local residents protested and, in the chaotic situation, US troops fired on the crowd, killing seventeen people.[44] Tensions continued to mount and on 31 March 2004, four American contractors of Blackwater Security Consulting were killed in Fallujah. Their bodies were burned and dragged through the streets by hundreds of Iraqis. Two of the corpses were suspended from a bridge over the Euphrates River. Subsequently, Falluja emerged as a center of anti-American resistance.

Seeking revenge, the United States launched operation "Desert Scorpion," a multi-prong campaign that sought to attack insurgents, strengthen ties with local leaders, and expand the local police force.[45] On 4 April the US marines laid siege to the city. As Phillips describes, the "marines erected earth barricades and cordoned off the city." In response, the citizens of Fallujah rallied to the cause of the insurgents, fighting Coalition forces street-by-street, house-by-house.[46]

Fallujah became a rallying point for the insurgency. Casualties mounted on both sides, with hundreds of Iraqis (many of who were noncombatants) dead, thousands wounded, and even more without water, electricity or food supplies.[47] Coincident with the three-week long siege by US Marines of Fallujah and the release of information on the torture and humiliation of Iraqi prisoners at Abu Ghraib, in April 2004 insurgents adopted a new tactic to undermine the resolve of the Coalition: abductions and video-taped executions.[48] On 12 April Italian security guard Fabrizio Quattrocchi and three other Italians were abducted in the area west of Baghdad. Quattrocchi was a former soldier in the Italian army and was believed to be in Iraq working for a US security company. The captors, identifying themselves as the Green Brigade, demanded that the Italian government withdraw its contingent of 3,000 soldiers from Iraq. Both Italian President Carlo Azeglio Ciampi and Prime Minister Silvio Berlusconi agreed that Italy would not yield to the abductor's demands. Signor Berlusconi declared in a statement that the Italian government would do everything in its power to secure

---

43 Kutler, "U.S. Military Fatalities," 543.
44 Phillips, *Losing Iraq*, 195.
45 Phillips, *Losing Iraq*, 196-198.
46 Phillips, *Losing Iraq*, 196.
47 Diamond, *Squandered Victory*, 234.
48 Phillips, *Losing Iraq*, 202.

the hostages' release, but would not give in to blackmail. Two days later Quattrochhi was executed.[49] His captors claimed to have killed the hostage because Berlusconi insisted that a troop withdrawal was not negotiable.

Soon thereafter another abduction garnered international headlines. This time it was an American, Nicholas Berg. Berg was in Iraq, apparently as a private citizen, seeking business opportunities. He owned a business, Prometheus Methods Tower Service, that inspected antennae on communications towers. He first went to Iraq in December 2003 on an 'exploratory' mission. He later returned to Iraq on March 14, 2004 with the intention of returning to the United States on March 30 to attend a friend's wedding. However, during this period he was jailed by US authorities in Iraq. After his release, Berg contacted his parents on four consecutive days, April 6 through 9. Berg contacted his parents, indicating that he had been arrested by Iraqi officials. He was then released and captured by the Abu-Mus'ab al-Zarqawi group, under the direction of Abu Musab al-Zarqawi.[50]

In a videotape posted on the Muqtada al-Ansar web site, a man identified as al-Zarqawi beheads a screaming Berg after reading a lenghty statement vowing more slayings for the "satanic degradation" of Muslims in the Abu Ghraib prison. Another man on the videotaped explained: "For the mothers and wives of American soldiers, we tell you that we offered the US administration to exchange this hostage for some of the detainees in Abu Ghraib and they refused." He continued: "So we tell you that the dignity of the Muslim men and women in Abu Ghraib and others is not redeemed except by blood and souls. You will not receive anything from us but coffins after coffins ... slaughtered in this way."[51]

Following the statements of al-Zarqawi, politicians, the media, and the American public, not surprisingly, made a direct connection between the prisoner abuse scandal at Abu Ghraib and the beheading of Berg. Admittedly, spokespersons would announce, the practices at Abu Ghraib were shameful, but at least they weren't killed. Military analyst Dan Goure explains that "This is standard operating procedure for them. So much for the moral equivalency that some people are trying to draw—we put women's panties over prisoners' heads and they cut off people's heads."[52] Steve Dunleavy, in *The New York Post*, was forthright in his condemnation. He wrote: "You say your unspeakable act of barbarity was retaliation for what happened in Abu Ghraib prison. What an excuse: A bunch of untrained Americans hazed and humiliated Iraqi detainees." Dunleavy, noting the abductors wore masks, argued that this was a sign of cowardice and weakness. If fact, Dunleavey argued, the

---

49 The three other workers abducted with Quattrocchi—Salvatore Stefio, Maurizio Agliana, and Umberto Cupertino—were subsequently released.

50 Brian Whitaker and Luke Harding, "American beheaded in revenge for torture," *Guardian Unlimited*, May 12, 2004, www.guardian.co.uk/international/story/0,3604,1214758,00.html (August 12, 2005).

51 Niles Lathem, "U.S. hostage screams in horror as he is beheaded," *The New York Post*, May 12, 2004, p. 2.

52 Deborah Orin, "Bush vows justice vs. barbarians," *The New York Post*, May 12, 2004, p. 4.

execution of Berg was a sign of weakness. He elaborated: "You and your wimpy terrorist brothers around the world are getting your butts kicked.... So what do you do? You kill an American when his arms are tied." Finally, Dunleavy warned, "if you think this ultimate act of barbarity will weaken the resolve of the coalition—and the decent people of Iraq—you are wrong."[53] In agreement, Andrew Sullivan in *The New York Sun* editorialized that the insurgents were vile, alien to true Islam, pathetic and dumb. He wrote: "They think they terrify us by this? The gang murder of an unarmed, innocent civilian? And they think that it will add to the shame of Abu Ghraib, demoralize Americans still further, and prompt a withdrawal?" Sullivan continued: "In the midst of our own deserved self-criticism, we are suddenly reminded of the larger stakes, the wider war, why we are in Iraq in the first place. Americans do not in any way excuse Abu Ghraib, but also see that any sort of moral equivalence between our flawed democracy and Islamism's pathological hatred is obscene."[54] But why did the United States invade and occupy Iraq? What are the larger stakes? Sullivan fails to recognize that the insurgency—albeit fueled by Abu Ghraib—was in fact responding to a wider war, namely the war for the control of Iraq's political and economic future.

US government counter-terrorism officials, in fact, explained that the videotaped beheading appeared to be designed by al-Zarqawi to capitalize on the worldwide outrage over the prison scandal and to weaken the will of an increasingly anxious American public to support the war effort in Iraq. It was also noted that aside from being a 'standard' tactic of Al Qaeda, Islamic fighters in the Afghan war against the Soviet Union also made numerous videotapes of beheadings of captured Russian soldiers that were sent home to their families.[55]

In some respects, the execution of Berg diverted attention from the sustained political motivation of the abductions. As the initial kidnapping of Quattrocchi demonstrated, abductions were carried out as a means to affect larger objectives. Subsequent abductions would confirm these goals. On 22 June the body of Kim Sun-il was found dumped outside of Baghdad. He was beheaded, and his body was booby-trapped with explosives. Kim had been abducted in Fallujah days earlier by the Al-Tawhid wa al-Jihad group. An Arabic speaker, Kim had been in Iraq for over a year, working as a translator for a South Korean firm, Gana General Trading Company, supplying goods to the United States military. After his capture, the abductors gave South Korea 24 hours to cancel plans to send 3,000 additional troops to Iraq. At the time, South Korea had a force of 600 in Iraq. The increase in troop

---

53 Steve Dunleavy, "Ski masks won't save your cowardly butts," *The New York Post*, May 12, 2004, p. 3.

54 Andrew Sullivan, "Just how stupid is al-Qaeda?" *The New York Sun*, May 14, 2004, p. 11.

55 Lathem, "U.S. hostage."

personnel would have made South Korea the largest coalition partner in Iraq after the United States and the United Kingdom.[56]

After the abduction, a video was broadcast on the Arabic-language television network Al-Jazeera. Kim was shown seated in front of three men whose faces were covered with scarves; two men held rifles at Kim. The third abductor delivered an ultimatum to South Korea's government: "We ask the government of South Korea and the people of Korea to pull their forces out of Iraq and not to send additional forces. Otherwise, we will send this hostage's head back to them and, God willing, we will kill more of your troops in Iraq. And you have 24 hours, starting tonight."[57] Kim, speaking in English, cried out in the video broadcast: "Please get me out of here. I don't want to die.... Your life is important, but my life is important."[58]

In a video broadcast after the execution of Kim, the abductors announced: "To the South Korean citizens: We warned you. This is the result of your own doings. Enough lies, or cheatings. Your soldiers here are not for the sake of the Iraqis, but they are here for the cursed America." The abductors allegedly belonged to a group known as 'Unity and Jihad'.[59] However, South Korean President Roh Moo-hyun instructed his ministries to explain to the Iraq people that the Korean government was sending troops to Iraq to focus on reconstruction efforts and not to engage in hostile acts.[60] Statements such as this would become a familiar mantra of foreign governments. Indeed, the death of Kim early on symbolized the cleavages of Occupied Iraq. Foreign governments continued to maintain that the presence of their troops fell under the umbrella of humanitarian aid and assistance; they did not consider their actions as part of a broader occupation. Insurgents viewed the situation differently.

On 16 September American engineers Jack Hensley and Eugene Armstrong, and British engineer Kenneth Bigley were kidnapped by the Al-Tawhid wa al-Jihad Group. The three hostages, all of whom worked for a Middle East-based construction firm, were abducted from their house in Baghdad. In a videotape, the captors threatened to behead the three Westerners within two days if the women detained in two US-run prisons in Iraq were not released. At the time, American officials insisted that no Iraqi women were being held at the two prisons specified. Officials did, later, admit to holding two female scientists because of alleged connections to developing anthrax. On 20 September it was disclosed that Armstrong had been executed. The following day Hensley met the same fate. A report posted on an Islamic website

---

56   Sohn Jie-Ae and Caroline Faraj, "Pentagon: South Korean hostage beheaded," *CNN. com*, June 22, 2004, www.cnn.com/2004/WORLD/meast/06/22/iraq.hostage/index.html (August 12, 2005).

57   Sohn Jie-Ae, "Militants threaten to behead South Korean hostage," *CNN.com*, June 21, 2004, www.cnn.com/2004/WORLD/meast/06/20/iraq.hostage/index.html (August 12, 2005).

58   Sohn and Faraj, "South Korean."

59   Sohn and Faraj, "South Korean."

60   Sohn and Faraj, "South Korean."

claimed that the "youths of Tawhid and Jihad killed the second American hostage after the end of the deadline."[61]

Following the deaths of Hensley and Armstrong, British Prime Minister Tony Blair said that there was little he could do to secure the release of Bigley. Earlier in the week a video was released by the captors, appealing directly to Blair, saying that there was no one else who could save the hostage. In an interview Blair explained "My first reaction is the reaction of anyone, which is real sympathy for him and anger at how he's being held by these people, and, you know the hope, the earnest hope that despite all the difficulties we can do something." In contrast, two members of the Muslim Council of Britain indicated that they would continue to secure the release of Bigley. Daud Abdullah, assistant secretary-general of the Council, indicated that his organization did not represent the British government. He explained: "How do you make a British citizen bear the responsibility for mistakes made by Tony Blair and his government? Regardless of the scandal of these mistakes, we are not judges. As Muslims, we cannot make anyone bear the responsibility of someone else's actions."[62]

The ordeal of Bigley continued for two more weeks until 8 October when it was announced that he had been beheaded. On British television Blair said that he felt a "strong sense that the actions of these people, whether in Iraq or elsewhere, should not prevail over people like Ken Bigley, who after all only wanted to make Iraq and the world a better place." One of Bigley's brothers, however, accused Blair of having blood on his hands. Britain had refused to negotiate with the captors. The British Foreign Secretary, in denouncing the killing, explained that the government had exchanged messages with Bigley's captors to try to secure his release, but that the British government also remained firm in its policy of non-negotiation and of never paying ransom to captors.[63]

Neither Italy, South Korea, the United States, nor Britain acquiesced to the demands of the captors. The same cannot be said of the Philippine government. As such, the abduction of Angelo de la Cruz, a Filipino truck driver working in Iraq, provides an interesting counter-narrative.[64] On 7 July—prior to the abductions of Armstrong, Hensley, and Bigley—de la Cruz was abducted near Fallujah. His abductors were identified as Khaled-Bin-al-Walid, the military wing of Al-Tawhid wa al-Jihad Group and those responsible for the death of Kim and (later) Armstrong,

---

61 CBC World News, "American hostage beheaded: video," September 20, 2004, www.cbc.ca/story/world/national/2004/09/20/hostage040920.html; CBC World News, "Another U.S. hostage beheaded, says website," September 22, 2004, www.cbc.ca/story/world/national/2004/09/21/hostages040921.html (September 29, 2005).

62 CBC World News, "Blair says there's little he can do to save British hostage in Iraq," September 26, 2004, www.cbc.ca/story/world/nationa/2004/09/26/blair040926.html (September 29, 2005).

63 CBC World News, "Hostage's killing 'barbaric,' Blair says," October 8, 2004, www.cbc.ca/story/world/national/2004/10/08/bigley041008.html (September 29, 2005).

64 For an in-depth look at the abduction of Angelo de la Cruz, see James A. Tyner, *Iraq, Terror, and the Philippines' Will to War* (Boulder, CO: Rowman & Littlefield, 2005).

Hensley, and Bigley. Unless the Philippines removed all of its military and police personnel from Iraq within seventy-two hours, the abductors would behead the hostage. At the time of the abduction, the Philippines had reportedly a token force of 51 peacekeepers stationed in Iraq. However, approximately 6,000 contract workers were also deployed in Iraq and neighboring Kuwait.

Initially, neither Philippine President Gloria Macapagal-Arroyo nor her Defense Secretary Eduardo Ermita favored withdrawing the peacekeeping force. Negotiations were, however, underway to secure the release of de la Cruz. Foreign Affairs secretary Delia Albert, for example, indicated that a Filipino Middle East Preparedness Team—already on the ground in Iraq—was handling the situation. Philippine charge d'affaires to Iraq, Eric Endaya, had already established contact with the Iraqi insurgents.[65]

By Sunday, as the deadline was approaching, Philippine officials attempted to persuade the abductors that if they freed de la Cruz, the Philippines would pull out their troops by August 20. This was in fact the scheduled date of their return. Albert reaffirmed that the government's policy was to retain the humanitarian mission in Iraq. National Security Advisor Norberto Gonzales indicated also that the administration was in consultation with influential Muslim leaders in several countries to talk with de la Cruz's captors.

On Monday, it was reported that a nine-day reprieve had been secured for de la Cruz. The demands for full withdrawal of Philippine forces remained in effect. It was later reported that the government offered to pay a ransom to the abductors in exchange for de la Cruz's life. Details were withheld as to how much ransom was offered, and whether the Philippine government or some other source would pay the ransom. The abductors, however, apparently refused the money, indicating that they would hold to their initial demands.

The ordeal came to a conclusion on 15 July when it was announced that the Philippine government had negotiated for the release of de la Cruz. Albert, in a statement read on local television, explained, "The Philippine government has recalled the head of the Philippine humanitarian contingent in Iraq. He is leaving Iraq today with 10 members of the humanitarian contingent."[66] The statement, while met with considerable joy in the Philippines, caught many officials outside of the Philippines by surprise.

The governments of Singapore, Poland, Australia and the United States were quick to condemn the Philippine government. The consensus was that any attempt to negotiate with 'terrorists' would embolden them and contribute to increased attacks. Ruth Urry, assistant information officer of the US Embassy in Manila, issued a

---

65 Jowie Corpuz, "Race against time to free de la Cruz," *Manila Times*, July 10, 2004, www.manilatimes.net/national/2004/jul/10/yehey/top_stories/20040710top1.html (July 16, 2004).

66 Jowie Corpuz and Ma. Theresa Torres, "Rumsfeld scorns Manila's 'weakness,'" *Manila Times*, July 23, 2004, www.manilatimes.net/national/2004/jul/23/yehey/top_stories/20040723top2.html (July 26, 2004).

one-page statement indicating that the United States was dismayed by the action of the Philippine government. She said, "This decision sends the wrong signal."[67] US Ambassador to the Philippines Francis Ricciardone explained that "In a time of crisis an ally, a friend, helps a partner to be strong and that's what we are trying to do. The ambassador added that the Philippines government should not confuse its enemies with its friends.[68]

The Australia reaction was more vitriolic. Australia's foreign minister, Alexander Downer, called on the Philippine ambassador to Canberra, Cristina Ortega, to convey his "extreme disappointment." Downer branded the Macapagal-Arroyo administration's decision as "marshmallowlike."[69] Chris Kenny, spokesperson for Downer, added that the Australian government attempted to persuade the Macapagal-Arroyo administration to remain firm to its commitment. Kenny elaborates, "If countries give in to terrorists, it will only encourage them to kidnap more hostages in an attempt to change the foreign policies of countries. Australia could not and would never do that."[70] Singaporean officials concurred with Australia. Tony Tan, Singapore coordination minister for security and defense, said, "The Singapore government cannot and should never negotiate with terrorists. That would encourage more terrorists to take more of our people as hostages."[71] At the time, Singapore had a force of 33 soldiers stationed in Iraq.

Whether or not attributable to the 'weakness' of the Philippines, abductions did continue in Iraq. As indicated above, in September—just one month after the release of de la Cruz—Armstrong, Hensley, and Bigley were abducted. But even sooner than that, in early July two Bulgarian truck drivers were abducted by those responsible for the deaths of Berg and Kim, the Al-Tawhid wa al-Jihad Group. In a statement, the abductors demanded, first, that all Iraqi detainees were to be released, and, second, that the United States withdraw its troops. This did not come to pass and both Georgi Lazov and Ivaylo Kepov were beheaded. Later that month, on 19 July 2004, Wasif Ali Hassun, a US Marine of Lebanese origin, was abducted, allegedly by the Islamic Retaliation Movement.[72] He was later released. Enzo Bladoni, an Italian journalist abducted by the Khaled-Bin-al-Walid Brigades in August 2004, was not as fortunate. He was later killed.[73] Next, on 23 July seven

---

67 Jowie Corpuz, "Pullout dismays US, other allies," *Manila Times*, July 15, 2004, www.manilatimes.net/national/2004/jul/15/yehey/top_stories/20040715top2.html (August 4, 2004).

68 Karl Kaufman, "Ricciardone: US, RP remain allies," *Manila Times*, July 16, 2004, www.manilatimes.net/national/2004/jul/16/yehey/top_stories/20040716top3.html (July 16, 2004).

69 Jowie Corpuz, "Singapore joins RP bashers; DFA mum," *Manila Times*, August 4, 2004,   www.manilatimes.net/national/2004/aug/04/yehey/top_stories/20040804top2.html (August 5, 2004).

70 Corpuz, "Pullout dismays US."

71 Corpuz, "Singapore joins."

72 Haddad and Ghazi, "Iraqi resistance groups."

73 Haddad and Ghazi, "Iraqi resistance groups."

hostages—three Kenyans, three Indians, and an Egyptian working for a Kuwait company in Iraq—were abducted. Another kidnapping was reported on 25 July, this time a senior Egyptian diplomat. The abductors demanded that Egypt give up any plans to send security experts to support Iraq's government. Egypt, which declined to send military forces, had offered to train Iraqi policy and security personnel in Egypt.[74] And still, on 31 August, a videotape was posted on the Internet showing the execution of twelve Nepalese hostages. The first man was beheaded, the other eleven were shot in the head.[75] As the hostage-taking escalated, both France and Germany warned their citizens to leave and Russian and Ukranian companies began to withdraw their workers.

By May 2005 more than 200 foreigners had been kidnapped in Iraq; approximately one-third of the hostages were killed. This does not include the unknown hundreds of Iraqis who have been abducted.

Power and Violence in Occupied Iraq

If the war and occupation of Iraq is, in the words of Derek Gregory, a colonial present, then the insurgency constitutes an anti-colonial and, by extension, an anti-capitalist movement. And as the previous chapters indicate, the present insurgency is the violent manifestation of converging transnational practices guided by a neoliberal ideology.

What I have argued thus far is that the movement of workers and warriors to Iraq is a process of a disciplined and controlled insertion of bodies into broader economic strategies. Contract workers—whether serving as soldiers or laborers— are commodified to fulfil certain objectives. As David Harvey explains, "capital continuously strives to shape bodies to its own requirements, while at the same time internalizing within its modus operandi effects of shifting and endlessly open bodily desires, wants, needs, and social relations (sometimes overtly expressed as collective class, community, or identity-based struggles) on the part of the laborer."[76] Recall from Chapter 1 my usage of the term 'subject'. In so doing, I emphasize the constructedness of bodies—subjects—and thus highlight the inscription of particular discourses onto these bodies. All bodies are in a process of becoming, and not necessarily of their own choosing. Embodied subjectivity is an effect of discursive and non-discursive practices.

The subjugated body, for example that of the migrant contract worker inserted into the reconstruction of Iraq, is in part the result of an exercise of power. Philip Barker, however, raises an interesting question with respect to the exercise of power. He asks: "If we are all implicated in practices of domination and subjection, then

---

74 Associated Foreign Press, "Egyptian first envoy abducted in Baghdad," *Manila Times*, July 25, 2004, www.manilatimes.net/national/2004/jul/25/yehey/top_stories/20040725top3. html (July 26, 2004).

75 Haddad and Ghazi, "Iraqi resistance groups."

76 David Harvey, *Spaces of Hope* (Cambridge, MA: Blackwell, 2000), 115.

how is it possible to distinguish between different applications of these practices?" He continues, "Does this not leave us unable to distinguish between the dominating practices that a parent may apply to their child, the dominating practices of the riot police against demonstrators, and the dominating practices of torture/death camps?" Barker, consequently, questions how it is possible to distinguish between "practices of domination and subjection" and an "outright engagement in unmitigated violence."[77]

Here it is useful to distinguish between power and violence. Power is relational. It is exercised only over free subjects, with 'free' meaning individual or collective subjects who are faced with a field of possibilities in which several kinds of conduct, several ways of reacting and modes of behavior are available. For Foucault, power exists on two conditions, the first being that 'the other' is recognized and maintained by a subject who acts, and the second being that, faced with a relationship of power, a whole field of responses, reactions, results and possible interventions may open up. Power is thus not simply a matter of consent but it is also not a renunciation of freedom, or a transfer of rights.

An exercise of violence, conversely, is totalizing. When violence is applied to a body, subjugation is complete. It removes the possibilities for active subjects to reinscribe themselves; it removes the possibility for resistance.[78] To this end, Barker explains that "violence involves a direct application of force upon the body of the other, reducing every possibility for independent action. Violence is applied directly to a body, but more than this it is applied to a body which is not recognized as being in a 'relationship' that would allow it to act autonomously."[79]

The abduction of hostages is a form of political violence; it is not an act of resistance. While Angelo de la Cruz, for example, was held in captivity, he was in a position of powerlessness.[80] De la Cruz in effect became a subjugated body. But this does not mean that his body was not still inscribed nor used for 'productive' purposes. As such, we need to consider the fundamental relations that comprise hostage situations. Here, I follow John Griffiths' terminology. The takers, or holders, of hostages are termed 'captors' or 'hostage-takers'; those taken, 'captives' or 'hostages'; and the authorities who actions, policies, or opinions and threatened, the 'coerced'. This tripartite division follows from Griffiths' understanding of political hostage-taking as "that in which the well-being or lives of unwilling captives are

---

77 Philip Barker, *Michel Foucault: An Introduction* (Edinburgh: Edinburgh University Press, 1998), 37.

78 Michel Foucault, "The Subject and power," in *Power, Essential Works of Foucault, 1954-1984*, Volume 3, edited by P. Rabinow (New York: The New Press, 2000), 326-48; at 340.

79 Barker, *Foucault*, 38.

80 I am sensitive to the arguments that captivity does not always constitute complete powerlessness. As studies of slave systems reveal, even in the most abject of conditions human agency may still find space to resist, if only in the mind. For present purposes, however, I operate on the assumption that publically, hostages in Occupied Iraq are not in a position to reassert themselves to the violent practices exercised against them.

threatened by their captors in order to compel a government or state authority to act in a way which it would not otherwise necessarily have acted."[81] The most immediate coerced participants, in this case, include the Philippine government as well as other members of the Coalition of the Willing.

During de la Cruz's captivity, both the Philippine state, the Iraqi insurgents, and other participants attempted to inscribe their own discourses onto the captive body of de la Cruz. Although powerlessness himself, de la Cruz continued to be subjected to various interpretations and meanings; his body, in effect, continued to work, albeit for larger political purposes. Philippine Foreign Secretary Delia Albert, for example, issued a statement that the Philippines was participating in a humanitarian effort to reconstruct Iraq. Citing the work performed by Filipino contract workers, Albert noted that "In less than a year of deployment in Iraq, the Philippine humanitarian contingent [had] constructed 21 schools, 4 footbridges, 3 public health centers, 2 multipurpose halls, repaired 11 water treatment tanks, built 2 kilometers of sewerage systems and asphalted 12 kilometers of road. They conducted 43 medical and civic-action programs, treating more than 14,000 Iraqis and also distributed food and medicine." Angelo de la Cruz, she explained, was part of that effort. Albert explained that "He wishes nothing more than an honest day's wage to feed his wife and 8 young children. He committed no acts of violence against the Iraqi people, nor does he wish them ill."[82]

From the perspective of the captors, de la Cruz was not an individual. Rather, his body—the subject—of de la Cruz signified something else entirely. De la Cruz, for the abductors, embodied not just one person, nor one nation, but many nations: the Philippines, the United States, the Coalition. He, and the other captives in Occupied Iraq embodied the abstract concepts of modernity and capitalism. This is made clear in the demands made by the abductors of the seven hostages following de la Cruz's release. When three Kenyans, three Indians, and an Egyptian were abducted on 23 July 2004, the hostage-takers intended, in part, to compel company to stop activities in Iraq.[83] Also, however, the abductors threatened to behead the men if their countries did not withdraw their troops and citizens from Iraq. Ironically, none of the hostages' governments had been part of the Coalition. The abductors, however, warned that every Kuwaiti company dealing with Americans would "be dealt with as an American."[84]

Such statements have become standard fare in Occupied Iraq. Following the execution of Kim Sun-il, and similar to the death of Berg, the abductors explained

---

81   John C. Griffiths, *Hostage: The History, Facts & Reasoning Behind Hostage Taking* (London: André Deutsche, 2003), 21, 28.

82   Wire Report, "UN's Annan Joins Plea to Save Angelo," *Manila Times*, 11 July 2004, www.manilatimes.net/national/2004/jul/11/yehey/top_stories/20040711top2.html (16 July 2004).

83   Haddad and Ghazi, "Iraqi resistance groups."

84   Associated Foreign Press, "Iraqi hostage-takers seize 7 truck drivers," *Manila Times*, 23 July 2004, www.manilatimes.net/national/2004/jul/23/yehey/top_stories/20040723top3.html (August 4, 2004).

to the South Korean government in a videotape aired on Al-Jazeera that "Your army is not here for the sake of Iraqis but for the sake of cursed America."[85] Alessandro Cevese, an Italian senior foreign ministry official, said that Fabrizio Quattrochhi may have been killed because he was carrying a card issued by the provisional authority of the US-led coalition.[86] And Nobutaka Watanabe, a human rights activist who was abducted and subsequently released in Iraq explained: "Our captors said from the beginning that they didn't capture us because they hated us, but because our country sent troops to their country."[87] In short, the bodies of workers and warriors, from the perspective of the abductors, are re-scripted as the personification of an illegal and unjustified occupation of their homeland.

Tortured Bodies and the Coerced

As bodies upon which violence is exercised, migrant workers become empty signifiers, used as political resources in the larger arena of international geopolitics. For this to be effective, however, the violence must be *visible*. To this end, the violence exercised on the body of Kim Sun-il, Nicholas Berg, Angelo de la Cruz and the other hostages forms part of a ritual; their ordeals are, from a Foucauldian perspective, techniques of torture. To understand how violence operates, it is instructive to consider another of Foucault's conceptions, that of torture and public execution. To be sure, not all hostages have been killed; some have been released. And even of those executed, not all have been tortured in a narrow usage of that term. Nevertheless, as a heuristic device, this focuses attention on the spectacle that often surrounds hostage-taking situations.

Foucault declares that torture is a technique; it is not an extreme expression of lawless rage.[88] Of particular importance is Foucault's assertion that torture forms part of a ritual, one that includes two components. First, it must mark the victim: it is intended to physically (as in scarring) or symbolically (through the accompanying spectacle) brand the victim with guilt. In the case of abductions, however, I maintain that it is not the individual per se that is branded with guilt, but instead another (usually governmental) entity. Second, and related, the public torture and execution must be spectacular, it must be seen by all almost as its triumph.[89] Consequently, the public display of Nicholas Berg's execution, or the graphic images of twelve executed Nepalese hostages are directed for a public, and global, consumption.

---

85   James Brooke, "Seoul Says Killing Won't Alter Plans for Iraq," *The New York Times*, 23 June 2004, Section A, Column 4, pg. 11.

86   John Hooper, "Iraq Fallout Hostages' Return Muted by Doubts on Their Mission," *The Guardian*, 10 June 2004, pg. 13.

87   Gabrielle Kennedy, "Japanese Politics Unsettles Hostages," *The Australian*, 28 April 2004, pg. 10.

88   Michel Foucault, *Discipline and Punish: The Birth of the Prison*, trans. by Alan Sheridan (New York: Vintage Books, 1979), 33.

89   Foucault, *Discipline and Punish*, 34.

The horrific death of the Italian worker Fabrizio Quattrocchi exemplifies further the ritualistic torture of some hostages. Prior to his execution Quattrocchi was videotaped while being forced to dig his own grave. He was then hooded by his captors and pushed to his knees before being shot in the neck.[90]

The violence directed against these bodies is not a punishment of people, for their bodies—to the captors—signify other elements. During the ritualistic displays of torture and captivity, no apparent attempt is made by the abductors to modify the behavior of Berg or de la Cruz, for example. Rather, the exercise of power occurs at another level; it is a relationship between the hostage-takers and the coerced, and not between the hostage-takers and the captives.

Publically, the captors of foreign and Iraqi hostages have declared that their actions are in reprisal for atrocities committed by Coalition forces. These include the death and destruction wrought by continued campaigns of 'shock-and-awe' as well as the prisoner abuse incidents at Abu Graib. Here the insights of Wole Soyinka are instructive:

> There they were, the would-be liberators, dehumanizing their prisoners and evidently relishing the experience. At once the contention of power and dignity was bared—literally—in blistering images. The world was treated to the performance of power when it becomes suddenly accessible to the powerless in relation to the powerlessness: the result was a graphic demonstration of the undiscriminating arrogance of power, manifested in the personal and sexual humiliation of the weaker, those who were made so only by circumstances, not by any intrinsic qualities ... simply by circumstances that could be reversed at any moment.[91]

Soyinka continues:

> Those circumstances were indeed reversed, and with an even more sickening escalation of horror, as a hostage was offered up to the world as a sacrificial lamb. The gruesome beheading of a hapless hostage in the name of reprisal, carried out in a manner that was clearly orchestrated toward global consumption, leapfrogged the incontinence of the US Army reservists in its barbarity, its arrogance and intensity of visceral laceration.[92]

Soyinka maintains that the killing was not 'purely' an act of vengeance. Instead, it was staged deliberately for global consumption and, indeed, instruction. The spectacle of Berg's death could be read thus as another statement of power, one that was directed at the world.[93] Such an interpretation resonates with previous

---

90 The death of Quattrocchi does reflect also that hostages do not necessarily conform according to the ritualistic script of the abductors. Prior to his execution, Quattrocchi struggled to remove the hood, shouting "I'll show you how an Italian dies." This scene, significantly, was *not* aired on Al Jazeera.

91 Wole Soyinka, *Climate of Fear: The Quest for Dignity in a Dehumanized World* (New York: Random House, 2004), xix.

92 Soyinka, *Climate of Fear*, xix.

93 Soyinka, *Climate of Fear*, xix-xx.

studies of suicide terrorism. Following Robert Pape, terrorist acts, such as suicide bombings, serve two broad purposes: to gain supporters and to coerce opponents.[94] Consequently, Pape identifies three different forms of 'terrorism': demonstrative, destructive, and suicide. *Demonstrative* terrorism is as much political theater as it is violence. It is a means of gaining publicity, to make one's grievances known to outside parties. Acts of demonstrative terrorism may also facilitate recruitment and gain attention from third parties who might exert pressure on the opposition. And lastly, demonstrative terrorism may be employed to disrupt supply lines or economic systems. Examples of this form of terrorism may include the bombing of an oil line or railroad track. *Destructive* terrorism, conversely, is more aggressive in that it seeks to coerce opponents with threat of injury or death, as well as to mobilize support. However, in its attempt to inflict real harm on members of target audiences—such as the bombing of a market or train station—the perpetrators risk losing sympathy for their cause.

Suicide terrorism is considered the most aggressive form of terrorism, pursuing coercion even at the expense of angering not only the target community but neutral audiences as well. From Pape's perspective, suicide terrorism differs from destructive terrorism in that in the former, the attacker does not expect to survive the mission.[95] Suicide terrorism could be used for demonstrative purposes, or could be limited to targeted assassinations. However, the trend is for attacks that maximize the coercive leverage; in effect, suicide bombings increasingly exert heavier losses of life.

In his work on suicide terrorism, Pape discusses the strategic, social, and individual logic of suicide terrorism. For my present purpose, I am most interested in his ideas of strategic logic and how these may apply to the abduction of foreign workers. The strategic logic of terrorism is aimed at political coercion. Pape notes that the vast majority of suicide terrorist attacks are not isolated or random acts by individual fanatics, but rather—and similar to hostage-taking incidents—occur in clusters as part of larger campaigns by organized groups to achieve specific political goals. Historically, these goals have been nationalistic rather than religious. Indeed, according to Pape, every group mounting a suicide campaign over the past two decades has had as a major objective—or as its central objective—coercing a foreign state that has military forces in what the terrorists see as their homeland to take those forces out.[96] This describes well the resistance movement in Occupied Iraq.

How then, have the various governments—the coerced—responded to abductions? As indicated by the de la Cruz abduction, the Philippine government overtly negotiated with the abductors, pleading for the release of their citizen. Angelo de la Cruz was constructed as a worker performing humanitarian acts, helping to rebuild the war-torn Iraqi infrastructure. Other governments, including that of South

---

94 Robert A. Pape, *Dying to Win: The Strategic Logic of Suicide Terrorism* (New York: Random House, 2005), 9.

95 Pape, *Dying to Win*, 10.

96 Pape, *Dying to Win*, 21.

Korea, likewise portrayed their workers in a humanitarian light. Their subsequent policies, however, were very different from that of the Philippines.

During the ordeal of Kim Sun-il the South Korean government remained firm in its decision to deploy an additional 3,000 troops to northern Iraq. Foreign Minister Ban Ki-Moon told reporters of the Associated Foreign Press that his governments "decision to dispatch ... troops for the purpose of helping the Iraqi people to rehabilitate their economy remains unchanged." The foreign minister did, however, arrange for the Jordanian foreign minister, Marwan Muasher, to help seek the release of Kim. Indeed, the South Korean government utilized a number of diplomatic channels, including the assistance of allied Middle Eastern states, to secure Kim's release. Influential religious and tribal leaders were also contacted.

From the perspective of the South Korean government, though, the abduction was viewed as separate from the larger Coalition efforts. Vice Foreign Minister Choi Young-jin explained that "The Kim Sun-il case has nothing to do with our planned dispatch. We will do our best to secure his release." Nevertheless, Choi maintained that "There is no change to the spirit of the dispatch and our position that the deployment is for supporting reconstruction and rehabilitation."[97] Such was the message South Korea attempted to promote to Iraq. During a 10 minute videotape aired on Al-Jazeera, South Korea officials stressed that "The South Korean troops have been dispatched to Iraq to help promote peace and reconstruct the country." They cautioned, moreover, that if Kim was executed, it would deal a serious blow to ties between South Korea and Iraq.[98]

Additional efforts were carried out by academics. A group of ten Koreans who taught Arabic at the Hankuk University of Foreign Studies in South Korea—the university at which Kim majored in Arabic and graduated in 2003—wrote a petition for the immediate release of Kim. Professor Park Jong-pyung, direct of the Institute of Middle East Studies, explained: "We wrote about his [Kim's] important role in Iraq, which is for the reconstruction and peace of Iraq. We explained why he went to Iraq and appealed in a peaceful manner to continue Korea's amicable relationship with Middle East countries." The petition opened with a quote from the Koran: "No one can harm anybody." The petition further explained that Kim was not a soldier but instead an interpreter who was working to help the Iraqi people. Significantly, the petition did not mention the deployment of Korean troops, thus continuing the government's separation of the hostage situation and the larger question of occupation.[99]

Attempts to secure the release of Kim proved futile and on 22 June the migrant worker was beheaded. In response, President Roh stressed that South Korean soldiers would remain in postwar Iraq. President Roh described the execution as a "crime

---

97 "South Korea Vows to Dispatch Troops to Iraq," Global News Wire, 21 June 2004.

98 "South Koreans Turn to Al-Jazeera in Bid to Save Kidnapped National in Iraq," BBC Monitoring Asia Pacific, 21 June 2004.

99 Chang Yeojean, "Professors of Arabic Join Appeals," *The Korean Herald*, 22 June 2004.

against humanity" and reaffirmed his government's position on the occupation. He explained that "We strongly condemn such terrorist acts and we will sternly combat terrorism in cooperation with the international community. We should never tolerate terror as a means to an end."[100] And despite mounting opposition to both the occupation and the actions of the government, a senior military official indicated that "It is shocking news, although we did not rule out the worst-case scenario. But it will not shake our will and stance. The government will prepare for the troop dispatch more thoroughly."[101] One report did indicate, however, that approximately 670 South Korean contract workers were ordered to leave Iraq before the end of June.[102]

Italy also remained steadfast in its position on Iraq following the death of Quattrocchi. In an open letter to the Quattrocchi family, President Carlo Azeglio Ciampi affirmed Italy's determination to stay the course and added, "The barbaric killing which has struck down your son in the fullness of his youth merely reinforces Italy's determination to block the path of hatred and to work for the realisation of peaceful co-existence in Iraq." Italian Prime Minister Silvio Berlusconi indicated that the man's death would not affect his country's "efforts for peace." He continued that "They have destroyed a life, they have not smashed either our values or our efforts for peace."[103]

The Japanese government has also remained firm in its commitment to not give in to terrorists' demands. When Shosei Koda, a 24-year old Japanese traveler was abducted by al-Zarqawi's group on 27 October 2004, Prime Minister Junichiro Koizumi said that Japan would not grant the kidnappers' demand for the withdrawal of its 550 troops from southern Iraq. Koizumi told reporters that "We cannot tolerate terrorism and we will not give in to terrorism."[104] That said, the Japanese government attempted to secure Koda's release through the assistance of approximately 25 foreign states.

Interestingly, Koda was not in Iraq as either a worker or a warrior. Rather, he entered Iraq as an 'adventure traveler'. Prior to his arrival in Iraq, Koda had traveled throughout New Zealand, Jordan, and Israel. He reportedly traveled to Iraq out of curiosity. His abductors, however, presumed Koda to be somehow linked to Japan's military presence in the country. As such, the kidnappers threatened to behead Koda if the Japanese government did not withdraw its forces. Upon learning of the death

---

100 Seo Hyun-jin, "Roh Condemns Killing of Hostage," *The Korean Herald*, 24 June 2004.

101 "Killing of S. Korean Hostage May Prompt Support for Troop Dispatch," Xinhua News Agency, 23 June 2004.

102 "Foreign Workers Leaving Iraq After Killing of South Korean Hostage," Xinhua News Agency, 23 June 2004.

103 Paddy Agnew, "Italians Show United Front After Killing of Hostage," *The Irish Times*, 16 April 2004, World News Section, pg. 13.

104 Anthony Faiola, "Japan Appeals for Release of Hostage in Iraq," *The Washington Post*, 28 October 2004, Section A21.

of Koda—his decapitated body was found two days later—Prime Minister Koizumi reaffirmed that "We cannot lose to terrorism; we must not yield to brute force."[105]

The coerced includes not only the targeted governments, but also members of the larger coalition. As indicated above, whereas the Philippines government was chastised by foreign governments, including the United States, Australia, and Singapore, for its capitulation to the abductors, other governments were praised for their actions. After the death of Kim, Australian Foreign Minister Alexander Downer said that "The South Korean government is showing enormous courage in keeping with its plan to deploy troops to Iraq."[106]

US President George Bush likewise condemned the beheading but expressed his confidence that the South Korean government would remain firm in its commitment to deploy the additional troops. Bush explained: "I would hope that President Roh would understand that the free world cannot be intimidated by the brutal action of these barbaric people." He continued: "They're trying to shake our will and our confidence. They're trying to get us to withdraw from the world so that they can impose their dark vision on people."[107]

President Bush's statement, however, came amid speculations of an American effort to buttress its support of the occupation. According to South Korean officials, American military officers knew of the 17 July abduction of Kim at least three days prior to informing the South Korean government. Reportedly, officials of South Korea's Defense Ministry indicated that representatives of the US marines stationed in Falluja, where Kim was abducted, deliberately withheld the information pending the South Korean governments confirmation of its troop dispatch on 18 July. Were the kidnapping made public prior to this time, spokespersons for the ministry suggested that it would have been difficult for the government to push ahead with the already unpopular plan to dispatch additional troops.[108] Although unconfirmed, these speculations raise serious questions about the underlying political subjugation of hostages, not only by the abductors, but also by the coerced. Workers and warriors are routinely sacrificed for ulterior motives.

The American occupation of Iraq and the sacrifice of workers and warriors is part of a neoconservative re-making of the global map. The mounting death-toll of American and coalition soldiers, foreign workers, and indigenous Iraqis is primarily attributable to the promotion of a militarized neoliberal agenda, an agenda predicated on a business of war. This agenda, moreover, has deep roots in the American society. As Stephen Kinzer writes, "Throughout the twentieth century and into the beginning of the twenty-first, the United States repeatedly used its military power, and that of

---

105 Norimitsu Onishi, "Koizumi Vows No Japanese Withdrawal After Tourists' Beheading," *The New York Times*, 1 November 2004, Section A, Column 3, pg. 10.

106 "Beheading of Korean Hostage Highlights Asian Split of Opinion Over Iraq," *Channel News Asia*, 24 June 2004.

107 "Bush Condemns Beheading of S. Korean Man in Iraq," Japan Economic Newswire, 22 June 2004.

108 "US Accused of Covering Up Kidnapping," *The Korean Times*, 23 June 2004.

its clandestine services, to overthrow governments that refused to protect American interests. Each time, it cloaked its intervention in the rhetoric of national security and liberation. In most cases, however, it acted mainly for economic reasons—specifically, to establish, promote, and defend the right of Americans to do business around the world without interference."[109] Such is the cause on which workers, warriors, and hostages are called upon to make sacrifices.

The steadfast refusal of the present administration to realize the pitfalls of a modern-day colonial project and the frailties of an empire built on coerced alliances and brute force has resulted in thousands of needless deaths and hundreds of thousands of life-altering injuries. Most egregious is the shallowness of political rhetoric that capitalizes on the sacrifice of others. Accordingly, it is instructive to consider Bush's perspective, within the context of hostages, on freedom and sacrifice more closely.

Bush has professed an ethic of respect for innocent human life. He has routinely situated his presidency within a 'culture of life.' However, as Peter Singer asks, "Is it consistent for someone who holds Bush's views about the sanctity of human life to be the supreme commander of armed forces that use bombs and missiles in areas where civilians are sure to be killed?" Singer, in short, questions Bush's culture of life in the context of wars in Afghanistan and Iraq.

It is a truism to say that wars kill people. Who is killed, however, remains ill-defined, based on the practice of warfare. To this end, there is a long history of discussion surrounding the concept of 'just-war'. Ghazi-Walid Falah, Colin Flint and Virginie Mamadouh maintain that "Modern just-war theory forces on the state as the legitimate actor, in contrast to the premodern focus on individual actions for the sake of individual honor."[110] Originating from the writings of Augustine in the fifth century AD, and modified through the work of Hegel, Kant, and Grotius, the justness of war is based on the idea that peace is achievable through war. More concretely, was is at present considered just when it occurs in the defense of state sovereignty.

This accounts for the extreme lengths to which American officials attempted to link Al Qaeda with Saddam Hussein, and to play up the (mis)information that Iraq possessed weapons of mass destruction. Within a militarized society war may indeed constitute significant business endeavors; this does not obviate the need for claims of legitimacy and justness. As such, the militant neoliberal re-making of the Middle East must be seen as just, and not solely as an expansion of empire.

Members of the Bush administration routinely maintain that sacrifice is required. But whom is making the sacrifice, and for what reason? On 10 April 2006 Bush spoke at the Paul H. Nitze School of Advanced International Studies as the Johns Hopkins University. Bush explained that the occasion "marked the third anniversary

---

109 Stephen Kinzer, *Overthrow: America's Century of Regime Change from Hawaii to Iraq* (New York: Times Books, 2006), 3).

110 Ghazi-Walid Falah, Colin Flint, and Virginie Mamadouh, "Just War and Extraterritoriality: the Popular Geopolitics of the United States' War on Iraq as Reflected in Newspapers of the Arab World," *Annals of the Association of American Geographers* 96(2006): 142-164; at 143.

of a great moment in the history of freedom"—the liberation of Iraq. Bush further explained that "today, because America and a great coalition acted, the regime is no longer in power, is no longer sponsoring terrorists, is no longer destablizing the region, is no longer undermining the credibility of the United Nations, is no longer threatening the world." He stated, simply, that "Because we acted, 25 million Iraqis now taste freedom." Explaining that the decision to "remove Saddam Hussein" was a difficult—though correct—decision, Bush maintained that America is now "doing its part to help the Iraqis build a democracy." And as for the Iraqis, Bush noted that "many have given their lives in the battle for freedom for their country. And by their courage and sacrifice, the Iraqi soldiers and civilians have shown that they want to live in freedom—and they're not going to let the terrorists take away their opportunity to live in a free society."[111]

Throughout his speech, Bush emphasized the idea of freedom, democracy, choice, and sacrifice. And he briefly situated these concepts within the context of the insurgency and hostage-taking. Reiterating the difficult decision to go to war, Bush acknowledged that "we've seen many contradictory images that are difficult for Americans to reconcile. On the one hand, we have seen images of great hope—boys and girls back in school, and millions of Iraqis dipping their fingers in purple ink, or dancing in the streets, or celebrating their freedom. On the other hand, we have seen images of unimaginable despair—bombs destroying hospitals, and hostages bound and executed." Bush concludes that these contradictory images raise "the question in the minds of many Americans—which image will prevail?" For Bush, the image is clear: "I believe that freedom will prevail in Iraq. I believe moms and dads everywhere want their children to grow up in safety and freedom."[112]

But these are not simply images. They are actual lives being killed. And they are not contradictory—selective, yes, as well as ambiguous, but not contradictory. Consider more specifically Bush's perception of executed hostages as one of unimaginable despair. Despair for whom? The hostages' family? The American audience writ large? For Bush, hostages are viewed in the negative, as failures of American foreign policy because he sees no meaning in their deaths. They are not seen as sacrificing for their countries—as are soldiers—because they are seen as helpless. They are powerless. As such, it is not possible for Bush to acknowledge their sacrifice: The death of hostages does not conform with the political rhetoric of a just war. These deaths, unlike the deaths of soldiers, signify a form of political violence that stands in opposition to the invasion and occupation of Iraq.

We should certainly not rejoice in the death of hostages. But neither should we assume that their lives (and deaths) have no meanings. Indeed, a parallel may be drawn between Bush's perception of hostages—and his concern that other government's not acquiesce to the demands of captors—with his concerns at the

---

111 Office of the White House, "President Bush Discusses Global War on Terror," 10 April 2006, [http://www.whitehouse.gov/news/releases/2006/04/20060410-1.html] (19 April 2006).

112 Office of the White House, "President Bush."

start of the Iraqi war. Prior to the 'shock and awe' that was unleashed in 2003 on the country of Iraq, Rumsfeld informed Bush that Saddam Hussein and his sons were supposedly located in a particular building. Rumsfeld wanted permission to bomb the site. Bush later explained in an interview that he was hesitant at first because he worried that the first pictures coming out of Iraq would be a wounded grandchild of Saddam Hussein. He was concerned, in his words, "that the first images of the American attack would be death to young children."[113] Singer, however, suggests that the concern Bush expresses is not about the risk that American bombs might kill or wound children. It was, rather, that Bush was concerned about potential negative reactions to the war. In other words, Bush was willing to risk the killing and maiming of innocent Iraqis in an attempt to perhaps kill Saddam Hussein.[114]

Sacrifice is tricky business. Presidents and other officials may call upon people in war-time to make sacrifices, but these sacrifices must have meaning. Most significant, however, is that the meaning of sacrifice must conform with the broader (and constructed) meaning of war. Given that the execution of hostages constitutes an act of resistance through political violence for the abductors, such actions must play negatively in the rhetoric of the coerced. From this perspective, hostages symbolize the loss of state sovereignty and, accordingly, cannot figure prominently in political rhetoric. Hostages are not seen as sacrificing for their country. Consequently, not only are the lives of hostages sacrificed by the coerced, so too is the meaning of their ordeals.

A Dehumanizing Occupation

I began this book with the intent to understand how hostages in Occupied Iraq embody the juncture and disruptures of neoliberal discourses and transnational, globalist practices. From the preceding chapters it is clear that presence of workers and warriors in Iraq—and the subsequent subjugation of workers and warriors as hostages—constitutes the confluence of these practices. I have argued that Occupied Iraq represents a contested place, one where bodies become dehumanized subjects. Workers, warriors, and hostages literally embody the moral vacuity of our present global society.

The visible abductions and executions of hostages is in the main a response to foreign occupation; it is an extreme strategy for national liberation against states with troops that pose an immediate threat to control territory the insurgents view as their homeland. The insurgency is, in Foucault's terminology, a struggle against subjection—against the submission of subjectivity.[115] The Coalition, viewed from this perspective, is exercising power over the Iraqi people. We should not deny, therefore,

---

113 Quoted in James A. Tyner, *Iraq, Terror, and the Philippines' Will to War* (Boulder, CO: Rowman & Littlefield, 2005), 122.

114 Peter Singer, *The President of Good and Evil: The Ethics of George W. Bush* (New York: Dutton, 2004), 158.

115 Foucault, "Subject and power," 332.

the violence perpetrated against the Iraqi people on the part of the Coalition. From this standpoint, the insurgency is in the collective a response, it is an act of resistance. Coalition forces are exercising power, subjecting the Iraqi people to a particular re-making. Chapter three illustrated well this subjectivity of the Iraqi people. Through the forced incorporation of neoliberal practices, the overall realignment of political, economic structures into a system that is conducive to American interests, the Iraqis, in short, are being subjected to a process of Americanization. And a number of Iraqis are resisting this subjugation. As Diamond explains, the "indignation over occupation and domination by the West was not a new phenomenon but, rather, summoned up decades and indeed centuries of wounds. Because Bremer and his colleagues in the CPA and the Bush administration never grasped this history, they could not anticipate how viscerally much of Iraq would react to an extended occupation."[116]

However, not all groups are able to exercise power to same degree. As such, we are witness to different forms of resistance within Occupied Iraq. Some individuals and groups have employed the more 'conventional' form of guerrilla warfare. Still others have concentrated on effective political change through the parameters established by the US-led Coalition. And still others have redirected their resistance through political violence, such as abductions, suicide bombings, or other techniques.

Nearly five decades ago, Frantz Fanon wrote that "decolonization is always a violent phenomenon."[117] The insurgency is not innocent of its atrocities. In no way do I condone the innocent slaughter of civilians, the abduction and execution of Iraqis and foreigners, the bombing of markets, schools, and places of worship. But an abhorrence to one atrocity does not lead to acquiescence to another. As Walden Bello writes, "to those quick to condemn such tactics [e.g., car bombings and kidnappings], one must point out that the indiscriminate killing of an estimated ten thousand Iraqi civilians by US troops in the first year of the occupation and the targeting of civilians in the siege of Falluja occupy the same moral plane as the methods used by the Iraqi and Islamic resistance."[118] Rather, one is sensitive to the spiral of de-humanism that is Occupied Iraq.

---

116 Diamond, *Squandered Victory*, 301.

117 Fanon, *Wretched of the Earth*, 35.

118 Walden Bello, *Dilemmas of Domination: The Unmaking of the American Empire* (New York: Metropolitan Books, 2005), 60.

Chapter 5

The Place of War

Under a pretext of a global war on terrorism, the administration of President George W. Bush embarked on a massive attempt to remake the political space of the Middle East. Situated within a rhetoric of 'security' and 'preemption' the Bush administration constructed the *causus belli* for an invasion and occupation of Iraq. Military victory was never in doubt. Less clear was the future of post-invasion Iraq. Subsequent events however leave little doubt that the military overthrow of Saddam Hussein was motivated in large part by a neoliberal desire to expand American capital interests in the Middle East, to secure oil and oil transportation routes in Central Asia, and to 'contain' the capitalist growth of competitors such as China. The Global War on Terror, in effect, masked the 'business' of war and occupation. Corporations such as Halliburton, Kellogg, Brown & Root, Blackwater, and Custer Battles have garnered substantial profits from the occupation. Indeed, by April 2004 Blackwater had more than 450 contractors in Iraq and millions of dollars worth of contracts in its coffers; notable among these contracts was the US$21 million Blackwater received for guarding Paul Bremer.[1]

A substantial amount of work has addressed the increased privatization of war, the growth of private military firms/corporations; and the presence of these PMFs/PMCs in Occupied Iraq. Decidedly less work has considered the other side of the coin, namely the increased privatization of the reconstruction effort and the use of foreign migrant contract workers. Numerous governments, from Bulgaria to Kenya, Nepal to the Philippines, attempted to capitalize on the destruction of Iraqi wrought by three major wars fought in the last twenty years and ten years worth of economic sanctions.

The United States as such is not alone in its profiteering of war. The Philippines, for example, was an early and eager participant in the Coalition of the Willing. Through the deployment of contract workers, the Philippine government expected to increase its revenues through the remittances of migrants. The Philippines' attempt to provide workers was part of a governmental effort to facilitate the deployment of transnational laborers and conformed readily with the long-standing neoliberal trend of privatized overseas employment in the Philippines. In George Orwell's *Nineteen Eighty-Four* wars were waged for the purpose of conducting future wars. The aim of warfare was not to win—for wars had become unwinnable—but instead to use the products of the war machine without raising the general standard of living. War had

---

1   Fred Rosen, *Corporate Warriors: The Rise of Privatized Military Industry* (Ithaca: Cornell University Press, 2003), 35.

become a business in itself. War was essential to facilitate the continued production, circulation, and consumption of war-related materials. As Orwell described, war "is a convenient way of expending labor power without producing anything that can be consumed." He continues that "It does not matter whether the war is actually happening, and, since no decisive victory is possible, it does not matter whether the war is going well or badly. All that is needed is that a state of war should exist." Ironically, Orwell also noted that in this state of permanent war, the Third World would serve as a surplus of excess labor. Orwell noted that, yes, these territories contained valuable minerals, but above all, they contained a bottomless reserve of cheap labor. He writes: "The inhabitants of these areas, reduced more or less openly to the status of slaves, pass continually from conqueror to conqueror, and are expended like so much coal or oil in the race to turn out more armaments, to capture more territory, to control more labor power, to turn out more armaments, to capture more territory, and so on indefinitely." The Philippines, among other states, has assumed such a position in the military neoliberal occupation of Iraq.

Place and Occupied Iraq

These two transnational systems—the privatization of the military and the use of contract workers—are thus part-and-parcel of a broader neoliberal agenda and quest for capital accumulation. These come together, violently, in the occupation and reconstruction of, as well as the contestation for, Occupied Iraq. Indeed, it is through a confluence of these transnational practices that the 'place' of Occupied Iraq is re-made and re-articulated. Place, in this sense, refers to the 'scale of everyday life'. As such, it entails two familiar ideas. On the one hand, it is a distinct point on the earth's surface. On the other hand, it is immediately local. As Noel Castree and colleagues write, "places are the product of myriad human practices." But places also are "the sedimentation of these practices over time," a process that "lends ... places their distinctiveness."[2]

When considering the place of Occupied Iraq, therefore, it is necessary to re-consider the meaning of place within a context of warfare. Places are normally conceived as self-contained entities, with a focus on internal relations. In short, the question is what goes on inside. Such a conception, however, presents two problems. First, this implies a rigid boundary between the 'local' and the 'non-local'. Second, it is obvious that places are not only interconnected, but are also interdependent. Castree and his co-writers explain that "It is no longer correct to assume that what is geographically juxtaposed is more important that what is geographically separated. Instead, geographical presence and absence co-mingle in varying ways." Consequently, these writers "see places less as bounded areas within a larger space

---

2   Noel Castree, Neil M. Coe, Kevin Ward, and Michael Samers, *Spaces of Work: Global Capitalism and the Geographies of Labour* (London: Sage Publications, 2004), 64.

of interconnections and interdependencies, but as the *meeting-place* of bundles of 'local' and 'non-local' events, processes and institutions."[3]

To this end, Castree and his co-authors identity three salient implications for such a conception of place. First, it shows that it is misguided to look solely 'within' a place if we are to understand what locally situated workers, firms and institutions are up to and what the effects of their actions are.[4] Consequently, any interpretation of the insurgency as well as the specific act of hostage-taking in Iraq must be approached from the standpoint of transnational interactions. Occupied Iraq is indeed a meeting place of American geopolitical interests, business decisions of private military firms, and private labor recruiters. Occupied Iraq is also the meeting-place of innumerable ideologies, all vying for the discursive and material control of localized governments, firms, and institutions.

Second, places may simultaneously represent homogenizing processes while at the same time retaining a sense of uniqueness. Again, to quote Castree and his co-writers, "Transnational interconnections, whereby commodities, people, information and images criss-cross the globe, may indeed be linking more places then ever. This fact of *different* places becoming swept-up in *similar* economic, political and cultural flows is ... one thing the concept of globalization directs our attention to." However, these writers also note that this "does not necessarily mean that places worldwide are becoming more alike." Indeed, "Places 'internalize' these processes in distinctive ways, which is why place interconnection does not imply increasing homogeneity among places."[5] This has profound implications for our understanding of the business of war in Occupied Iraq. Whereas it is possible to see American military intervention in Iraq as but the latest episode of a longer agenda of capitalist expansion, it may not be possible—nor desirable—to draw definitive lessons or comparisons from previous episodes. Iraq is *not* Vietnam; nor is it Kosovo or Somalia. The transnational practices that have converged in Iraq—and with these, the presence of workers, warriors, and insurgents—constitute both a continuation and a departure from those practices enacted in other places at other times. It is this "complex intermingling of myriad local and extra-local connections [that] determine the nature of most places" that has been lost in most discussions of Iraq. The attempts by Bremer's Coalition Provisional Authority, for example, were an attempt to radically re-configure the political, social, and economic contours of Iraq. His directives however demonstrated a geopolitical arrogance and ignorance of the country.

Lastly, and following Castree et al., "seeing places as open and porous allows us to understand why they are not simply different from one another but *unevenly and causally related.*"[6] This brings us back to my earlier discussion of the exercise of power and the attempted state-building conducted by the Coalition. Occupied

---

3   Castree, et al., *Spaces of Work*, 66.
4   Castree et al., *Spaces of Work*, 66.
5   Castree et al., *Spaces of Work*, 67-68.
6   Castree et al., *Spaces of Work*, 68.

Iraq is the focal point, the point at which all transnational practices, all circuits of capital and labor, converge and interact. In terms of humanity, however, it is a site of struggle. Obscured behind the estimates of 'collateral damage' and 'acceptable losses' are men, women, and children who are caught in the geopolitical webs of a militant neoliberalism.

The Making and Meaning of Hostages

This study has concentrated on the confluence of two transnational practices, both intimately connected with broader structural and ideological trends, namely neoliberalism, neoconservativism, and militarism. In *The Business of War*, though, I focus specifically, though not exclusively, on the actions of both the United States and the Philippines. My justification is simple. These two states constitute the largest providers of warriors and workers, respectively. The US is the provider of warriors (whether through private firms or the 'regular' military) and the Philippines is the largest provider of contract workers. Indeed, many of the Filipino workers were sub-contracted by American firms for the reconstruction of Iraq. By focusing on these two trends—the transnational movement of warriors and workers—I subsequently consider the making of hostages within Occupied Iraq.

Since the war began on 20 March 2006 to the present-day (April 2006), over 425 foreigners have been abducted in Occupied Iraq. Countless Iraqis have likewise been kidnapped. How many of these people have been killed as opposed to released is not exactly known.

John Griffiths, in his historical study of hostage-taking, concludes that "A combination of developments has diminished the value placed on human life and, therefore, of the bargaining leverage of threatening to kill a hostage. The multitudinous of mankind ... and the rash of lethal conflicts across the globe, have diminished the perceived importance of any one individual life other than our own and that of our nearest and dearest."[7] Griffiths' point is that in our society we have become numbed to the plight of hostages. Certainly, with few exceptions—such as the release of American Jill Carroll—the taking of hostages in Occupied Iraq has received far less media attention than one might expect. Indeed, compared to the extensive media coverage of US Vice President Dick Cheney's accidental shooting of Texas-based lawyer Harry Wittingham, there has been no sustained coverage of hostages related to the occupation of Iraq. Griffiths contends that, regarding hostage-taking in general, that "The media, with its daily diet of unimaginative fictional mayhem and the supposedly objective reporting of all-too-real mayhem in the world at large, has fed an appetite for the ever more shocking and bloody that has dulled our sensibilities to such a degree that horrors have to be even more horrific to awaken them. They can thus be more easily traded off for political reasons without too many pangs of national conscience." He concludes: "Our sympathy for the plight of others

---

7   John C. Griffiths, *Hostage: The History, Facts & Reasoning Behind Hostage Taking* (London: André Deutsche, 2003), 202-03.

decreases in length and strength proportionately to the increasing demands made on it."[8] Simply put, unless hostages in Occupied Iraq are notable for their horrific or unexpected outcome—such as the grisly execution by decapitation—hostage-taking in itself is not spectacular enough to garner attention. Coupled with short attention spans, the public at large is also less likely to follow an abduction case that last for weeks (if not months). Indeed, the personal web-sites established by friends and families of hostages in Occupied Iraq write of their dismay from a lack of attention to the plight of their loved ones. Consider the plight of Roy Hallums and his family. Hallums, a 56-year-old contractor, was abducted on 1 November 2004 while working for a Saudi Arabian catering company that serviced the Iraqi army. He was kidnapped along with five other men, including Filipino workmate Roberto Tarongoy. Six months into his captivity, Hallums' ex-wife of 30 years, Susan Hallums, was interviewed by the Australian Broadcasting Corporation. During the interview Susan Hallums discussed her dismay and frustration with having no information on the condition of her former husband. Even more discouraging, however, was the lack of attention in general toward Roy Hallums' ordeal. Kerry O'Brien, who interviewed Susan, asked: "Have you and the family had much sense of American support, American public sympathy behind you? Do you think they are even particularly aware of the case six months on?" Susan Hallums replied: "Uhm, I mean, even friends of ours that were close to us that have called me in the last few weeks that have just found out about it because it hasn't—it's sort of silenced and it's been nothing like when other countries have had hostages and there's huge protest. Here it is pretty quiet."[9]

Griffiths suggests that political hostage-taking was the crime of the hopeful. It was based on the belief that in return for the lives of hostages, meaningful concessions would be made sufficiently often by the coerced to make the risks more worthwhile. According to Griffiths, writing in general, this, however, has not happened. Consequently, there has been a rise in other forms of political violence, such as suicide bombings. Griffiths maintains that a transformation has occurred whereby the militant—in an attempt to attract the attention of the media or to influence a stubborn government—has changed from wishing to influence his or her enemy, to wishing only to destroy that enemy. As such, he concludes that the importance of hostage-taking will continue to diminish.[10]

Such an assessment may hold true for hostage-taking in general, but it does not appear to conform with events in Occupied Iraq. Indeed, the early months of 2006 have witnessed an increased trend of abductions. Why this is the case is open to speculation. Earlier I argued that hostages are empty signifiers, re-inscribed with

8   Griffiths, *Hostage*, 203.

9   Kerry O'Brien, "Kidnapped US Man's Family Awaits News," *Australian Broadcasting Corporation*, 5 May 2005 [http://www.abc.net.au/7.30/content] (4 April 2006). Readers may also visit the web-page created by Roy Hallum's family at http://royhallums.4t.com. Hallums was released in September 2005 after spending 10 months in captivity.

10   Griffiths, *Hostage*, 204.

various political messages by both the captors and the coerced. Given the relative silence surrounding the plight of hostages in the countries of the coerced, one wonders as to the overall efficacy of this technique. Hostage-taking may result in modest material gains, perhaps the payment of a ransom. But in Occupied Iraq, one particular motivation of the abduction of (especially) foreign hostages is to make a spectacular case for attention. Historically, as Griffiths acknowledges, hostage-taking has achieved greater awareness of the cause on behalf of which the abduction was undertaken. Consequently, the public-at-large was made more sharply aware of individual instances of injustice, oppression, or thwarted legitimate aspirations to nation-hood.[11]

But this does not hold in Occupied Iraq. The American public, as a case in point, is not in a position to view the insurgency as an instance of injustice or oppression. Indeed, the plight of hostages does not conform with the present administration's construction of the invasion and occupation of Iraq as a pivotal battle in the larger War on Terror. This may, in part, account for the relative silence of hostage-taking in the United States.[12] Moreover, any legitimacy that the insurgents may have acquired is lost when the world witnesses the ritual execution of captive workers. Consequently, one must reconsider the practice of abductions in Occupied Iraq. It may very well be that these kidnappings are designed not to influence western audiences, but instead to reestablish a position of power within the minds of Iraqis and other Arab and Muslim populations.

Dissonant Imaginations

Derek Gregory writes that "the modern world is marked by spacings of connection, which are worked by transnational capital circuits and commodity chains, by global flows of information and images, and by geopolitical alignments and military dispositions." These connections, he suggests, "have their own uneven geographies—they do not produce a single, smooth surface—and they are made intelligible through their own imaginative geographies." However, the "modern world is also marked by spacings of disjuncture between the same and the other that are installed through the same or parallel economic, cultural, and political networks but articulated by countervailing imaginative geographies that give them different force and sanction." These then are, in Gregory's term, *mappings of connective dissonance*, and are immediately recognizable within Occupied Iraq.[13]

---

11 Griffiths, *Hostage*, 203.

12 It bears mentioning that most books published thus far on the occupation and the Iraqi insurgency make only passing reference to the abductions. See, for example, David L. Phillips, *Losing Iraq: Inside the Postwar Reconstruction Fiasco* (New York: Westview Press, 2005) and Larry Diamond, *Squandered Victory: The American Occupation and the Bungled Effort to Bring Democracy to Iraq* (New York: Henry Holt and Company, 2005).

13 Derek Gregory, *The Colonial Present: Afghanistan, Palestine, Iraq* (Malden, MA: Blackwell, 2004), 255-56.

Workers, warriors, and hostages are literally trapped within these conflicting and contested imaginations of what Iraq is to become. Transnational circuits of migrants and militants, capital and commodities, coalesce on the streets and plazas of Baghdad, Fallujah, Nasiriyah, Mosul, and beyond. Occupied Iraq has become both a space of connection and a space of disjuncture; it constitutes a jarring clash not of civilizations, but of ideologies and material practices. But even more so it has come to signify a space of dehumanism, wrought by a militant neoliberal and neoconservative agenda to restructure the world predicated on free markets and private property. As Gregory explains, the war on terror—coupled with the wars in Iraq, Afghanistan, and elsewhere—constitutes an attempt to establish a new global narrative in which the power to narrate is vested in a particular constellation of power and knowledge, located primarily but not exclusively in the United States.[14] Such is the locus of geographical imaginations.

George Orwell is considered the 'wintry conscience' of his generation. According to Jeffrey Meyers, Orwell "fought for social justice and believed that it was essential to have both personal and political integrity."[15] In Orwell's writings, he attempted to convey a sense of humanity within de-humanizing worlds. His task was to articulate alternative imaginations, different mappings of possibilities, to open new vistas for social justice. In our current age of neoliberalism, neoconservatism, and militarism, we must continue the task set forth by Orwell. We owe nothing less to our future generations.

---

14 Gregory, *Colonial Present*, 16.
15 Jeffrey Meyers, *Orwell: Wintry Conscience of a Generation* (New York: W.W. Norton and Company, 2000), 325.

Bibliography

Abbey, Edward, *Abbey's Road* (New York: Penguin Books, 1972).

Asia, Maruja M.B., "The Overseas Employment Program Policy," in *Philippine Labour Migration: Impact and Policy*, edited by G. Battistella and A. Paganoni (Quezon City: Scalabrini Migration Center, 1992), 68-112.

Bacevich, Andrew J., *The New American Militarism: How Americans are Seduced by War* (Oxford: Oxford University Press, 2005).

Ball, Rochelle and Piper, Nicola, "Globalisation and Regulation of Citizenship—Filipino Migrant Workers in Japan," *Political Geography* 21 (2002): 1013-1034.

Baradat, Leon P., *Political Ideologies: Their Origins and Impact*, 7th ed. (Upper Saddle River, NJ: Prentice Hall, 2000).

Barker, Philip, *Michel Foucault: An Introduction* (Edinburgh: Edinburgh University Press, 1998).

Basch, Linda, Glick Shiller, Nina, and Blanc, Cristina Szanton, *Nations Unbound: Transnational Projects, Postcolonial Predicaments, and Deterritorialized Nation-States* (Amsterdam: Gordon and Breach, 1994).

BBC, www.bbc.co.uk.

Bello, Walden, *Dilemmas of Domination: The Unmaking of the American Empire* (New York: Henry Holt and Company, 2005).

Bello, Walden, Kinley, D., and Elinson, E., *Development Debacle: The World Bank in the Philippines* (San Francisco: Institute for Food and Development Policy, 1982).

Belsey, Catherine, *Poststructuralism: A Very Short Introduction* (Oxford: Oxford University Press, 2002).

Bennis, Phyllis, *Before & After: US Foreign Policy and the War on Terrorism* (New York: Olive Branch Press, 2003).

Boal, Iain, Clark, T.J., Matthews, Joseph, and Watts, Michael, *Afflicted Powers: Capital and Spectacle in a New Age of War* (New York: Verso, 2005).

Burbach, Roger and Tarbell, Jim, *Imperial Overstretch: George W. Bush and the Hubris of Empire* (New York: Zed Books, 2004).

Cainkar, Louise, "The Impact of the September 11 Attacks on Arab and Muslim Communities in the United States," in *The Maze of Fear: Security and Migration after 9/11*, edited by John Tirman (New York: The New Press, 2004).

Callinicos, Alex, *The New Mandarins of American Power: The Bush Administration's Plans for the World* (Cambridge, UK: Polity Press, 2003).

Cameron, Fraser, *US Foreign Policy After the Cold War: Global Hegemon or Reluctant Sheriff?* (New York: Routledge, 2002).

Carnoy, Michael and Castells, Manuel, "Globalization, the knowledge society, and the network state: Poulantzas at the Millennium," *Global Networks* 1 (2001): 1-18.

Casco, Richard, *Full Disclosure Policy: A Philosophical Orientation* (Manila: Philippine Overseas Employment Administration, 1995).

————. *Managing International Labour Migration and the Framework for the Deregulation of the POEA* (Manila: Philippine Overseas Employment Administration, 1997).

Castree, Noel, Coe, Neil M., Ward, Kevin, and Samers, Michael, *Spaces of Work: Global Capitalism and the Geographies of Labour* (London: Sage Publications, 2004).

Césaire, Aimé, *Discourse on Colonialism*, trans. By Joan Pinkham (New York: Monthly Review Press, 2000 [1955]).

Chatterjee, Pratap, *Iraq, Inc.: A Profitable Occupation* (New York: Seven Stories Press, 2004).

Chin, Christine B.N., *In Service and Servitude: Female Foreign Domestic Workers and the Malaysian 'Modernity' Project* (New York: Columbia University Press, 1998).

Christian Science Monitor, www.csmonitor.com.

Clark, William R. *Petrodollar Warfare: Oil, Iraq and the Future of the Dollar* (Gabriola Island: New Society, 2005).

CNN, www.cnn.com.

Cresswell, Tim, "Falling Down: Resistance as Diagnostic," in *Entanglements of Power: Geographies of Domination/Resistance*, edited by Joanne P. Sharp, Paul Routledge, Chris Philo, and Ronan Paddison (London: Routledge, 2000), 256-268.

Daalder, Ivo H., and Lindsay, James M., *America Unbound: The Bush Revolution in Foreign Policy* (Washington, D.C.: Brookings Institute Press, 2003).

D'Arcus, Bruce, "Globalization and Protest: Seattle and Beyond," in *WorldMinds: Geographical Perspectives on 100 Problems*, edited by Donald G. Janelle, Barney Warf, and Kathy Hansen (Boston: Kluwer, 2004), 21-24.

Diamond, Larry, *Squandered Victory: The American Occupation and the Bungled Effort to Bring Democracy to Iraq* (New York: Henry Holt and Company, 2005).

Dicken, Peter, *Global Shift: The Internationalization of Economic Activity*, 2nd ed. (New York: The Guilford Press, 1992).

Dinnerstein, Leonard, Nichols, Roger L., and Reimers, David M., *Natives and Strangers: A Multicultural History of Americans* (New York: Oxford University Press, 1996).

Dupont, Alan, "Transnational Violence in the Asia-Pacific: An Overview of Current Trends," in *Terrorism and Violence in Southeast Asia: Transnational Challenges to States and Regional Stability*, edited by Paul J. Smith (Armonk, NY: M.E. Sharpe, 2005), 3-18.

Edkins, Jenny, *Poststructuralism and International Relations: Bringing the Political Back In* (Boulder, CO: Lynn Rienner Publishers, 1999).

Engdahl, William. *A Century of War: Anglo-American Oil Politics and the New World Order*, revised edition (London: Pluto Press, 2004).

Everest, Larry, *Oil, Power, and Empire: Iraq and the U.S. Global Agenda* (Monroe, ME: Common Courage Press, 2003).

Falk, Richard, *The Great Terror War* (New York: Olive Branch Press, 2003).

Fanon, Frantz, *The Wretched of the Earth* (New York: Grove Press, 1963).

Flint, Colin and Falah, Ghazi-Walid, "How the United States Justified its War on Terrorism: Prime Morality and the Construction of a 'Just War,'" *Third World Quarterly* 25 (2004): 1379-1399.

Foucault, Michel, *The Archaeology of Knowledge and the Discourse on Language*, trans. A.M. Sheridan Smith (New York: Pantheon Books, 1972).

_____. *Discipline and Punish: The Birth of the Prison*, trans. A. Sheridan (New York: Vintage Books, 1979).

_____. "Questions on Geography," in *Power/Knowledge: Selected Interviews and Other Writings, 1972-1977*, edited by C. Gordon (New York: Pantheon Books, 1980), 63-77.

_____. "Two Lectures," in *Power/Knowledge: Selected Interviews and Other Writings, 1972-1977*, edited by C. Gordon (New York: Pantheon Books, 1980), 78-108.

_____. "The Subject and Power," in *Power: Essential Works of Foucault, 1954-1984*, Volume 3, trans. R. Hurley and others (New York: The New Press, 2000), 326-348.

_____. "Technologies of the Self," in *Ethics: Subjectivity and Truth, Essential Works of Foucault, 1954-1984*, Volume 1, edited by P. Rabinow (New York: The New Press, 2000).

Freedman, Lawrence and Karsh, Effraim, *The Gulf Conflict 1990-1991: Diplomacy and War in the New World Order* (Princeton, NJ: Princeton University Press, 1993).

Froehling, Oliver, "The Cyberspace 'War of Ink and Internet' in Chiapas, Mexico," *The Geographical Review* 87 (1997): 291-307.

Gaddis, John Lewis, *Strategies of Containment: A Critical Appraisal of Postwar American National Security Policy* (New York: Oxford University Press, 1982).

Guardian, www.guardian.co.uk.

Gerstle, Gary, "The Immigrant as Threat to American Security: A Historical Perspective," in *The Maze of Fear: Security and Migration after 9/11*, edited by John Tirman (New York: The New Press, 2004).

Giroux, Henry A., *The Terror of Neoliberalism* (Boulder, CO: Paradigm Publishers, 2004).

Gonzalez, III, Joaquin L., *Philippine Labour Migration: Critical Dimensions of Public Policy* (Singapore: Institute of Southeast Asian Studies, 1998).

Gordon, N., "On Visibility and Power: An Arendtian Corrective of Foucault," *Human Studies* 25 (2002), 125-145.

Gray, John, *Al Qaeda and What it Means to be Modern* (New York: The New Press, 2003).

Gregory, Derek, *The Colonial Present: Afghanistan, Palestine, Iraq* (Malden, MA: Blackwell, 2004).

Griffiths, John C., *Hostage: The History, Facts & Reasoning Behind Hostage Taking* (London: André Deutsche, 2003).

Gunaratna, Rohan, *Inside Al Qaeda: Global Network of Terror* (New York: Berkley Books, 2003).

Harding, Jim, *After Iraq: War, Imperialism, and Democracy* (London: Merlin Press, 2004).

Hartung, William D., "Military," in *Power Trip: U.S. Unilateralism and Global Strategy After September 11*, edited by John Feffer (New York: Seven Stories Press, 2003), 69-74.

Harvey, David, *The Limits to Capital* (Oxford: Basil Blackwell, 1982).

_____. *The Condition of Postmodernity* (Oxford: Basil Blackwell, 1989).

_____. *Spaces of Hope* (Cambridge, MA: Blackwell, 2000).

_____. *Spaces of Capital* (New York: Routledge, 2001).

_____. *The New Imperialism* (Oxford: Oxford University Press, 2003).

_____. *A Brief History of Neoliberalism* (Oxford: Oxford University Press, 2005).

Hawes, Gary, *The Philippine State and the Marcos Regime: The Politics of Export* (Ithaca, NY: Cornell University Press, 1987).

Howard, Russell D., "Understanding al Qaeda's Application of the New Terrorism— The Key to Victory in the Current Campaign," in *Terrorism and Counterterrorism: Understanding the New Security Environment*, edited by Russel D. Howard and Reid L. Sawyer (Guilford, CT: McGraw-Hill, 2004), 75-85.

Howard-Pitney, David, *The Afro-American Jeremiad: Appeals for Justice in America* (Philadelphia: Temple University Press, 1990).

Johnson, Chalmers, *Sorrows of Empire: Militarism, Secrecy, and the End of the Republic* (New York: Henry Holt and Company, 2004).

Jordan, Amos A., Taylor, William J., and Mazaar, Michael J., *American National Security*, 5th ed. (Baltimore: Johns Hopkins University Press, 1999).

Karnow, Stanley, *In Our Image: America's Empire in the Philippines* (New York: Random House, 1989).

Kellner, Douglas, *From 9/11 to Terror War: The Dangers of the Bush Legacy* (Lanham, MD: Rowman & Littlefield, 2003).

Kelly, Philip, F., *Landscapes of Globalization: Human Geographies of Economic Change in the Philippines* (London: Routledge, 2000).

Kinzer, Stephen, *All the Shah's Men: An American Coup and the Roots of Middle East Terror* (New York: John Wiley & Sons, 2003).

Klare, Michael T., *Blood and Oil: The Dangers and Consequences of America's Growing Petroleum Dependency* (New York: Metropolitan Books, 2004).

Kuruvilla, S., "Economic Development Strategies, Industrial Relations Policies, and Workplace IR/HR Practices in Southeast Asia," in *The Comparative Political Economy of Industrial Relations*, ed. K. Wever and L. Turner (Madison: Industrial Relations Research Association Series, University of Wisconsin, 1995), 115-150.

Lustick, Ian S., "America's War and Osama's Script," *The Arab World Geographer* 6 (2003): 24-26.

Manila Times, www.manilatimes.net.

Mann, Michael, *Incoherent Empire* (London: Verso, 2003).

McLaren, M.A., *Feminism, Foucault, and Embodied Subjectivity* (Albany: State University of New York Press, 2002).

Meyers, Jeffrey, *Orwell: Wintry Conscience of a Generation* (New York: W.W. Norton & Company, 2000).

Mitchell, Kathryn, *Crossing the Neoliberal Line: Pacific Rim Migration and the Metropolis* (Philadelphia: Temple University Press, 2004).

MSNBC, www.msnbc.com.

Nevins, Joseph, *A Not-So-Distant Horror: Mass Violence in East Timor* (Ithaca, NY: Cornell University Press, 2005).

New Yorker, www.newyorker.com.

Oakes, Timothy, "Place and the Paradox of Modernity," *Annals of the Association of American Geographers* 87 (1997): 509-531.

Office of the Philippine President, www.ops.gov.ph.

Office of the Press Secretary, www.whitehouse.gov.

Ong, Aihwa, *Flexible Citizenship: The Cultural Logics of Transnationality* (Durham, NC: Duke University Press, 1999).

Orwell, George, *A Collection of Essays* (New York: Harcourt, 1981).

_____. *Nineteen Eighty-Four* (New York: Plume, 1983 [1949]).

Pape, Robert, *Dying to Win: The Strategic Logic of Suicide Terrorism* (New York: Random House, 2005).

Parker, Robert E., *Flesh Peddlers and Warm Bodies: The Temporary Help Industry and its Workers* (New Brunswick, NJ: Rutgers University Press, 1994).

Peet, Richard, *Global Capitalism: Theories of Societal Development* (New York: Routledge, 1991).

Philippine Overseas Employment Administration, "Market Development: Seeking Purpose and Promise for Filipino Skills," *Overseas Employment Info Series* 1 (1988): 5-9.

_____. *Migrant Workers and Overseas Filipinos Act of 1995: Republic Act 8042 and its Implementing Rules and Regulations* (Manila: Department of Labor and Employment, 1996).

Philippine Star, www.philstar.com.

Phillips, David L., *Losing Iraq: Inside the Postwar Reconstruction Fiasco* (New York: Westview, 2005).

Prestowitz, Clyde, *Rogue Nation: American Unilateralism and the Failure of Good Intentions* (New York: Basic Books, 2003).

Ramonet, Ignacio, *Wars of the 21ˢᵗ Century: New Threats, New Fears*, trans. Julie Flanagan (New York: Ocean Press, 2004).

Reuber, "The Tale of the Just War—A Post-Structrualist Objection," *The Arab World Geographer* 6 (2003): 44-46.

Richardson, Louise, "Global Rebels: Terrorist Organizations as Trans-National Actors," in *Terrorism and Counterterrorism: Understanding the New Security Environment*, edited by Russel D. Howard and Reid L. Sawyer (Guilford, CT: McGraw-Hill, 2004), 67-73.

Robbins, James S., "Bin Laden's War," in *Terrorism and Counterterrorism: Understanding the New Security Environment*, edited by Russel D. Howard and Reid L. Sawyer (Guilford, CT: McGraw-Hill, 2004), 392-404.

Rosen, Fred, *Contract Warriors* (New York: Alpha, 2005).

Sarup, Madan, *An Introductory Guide to Post-Structuralism and Postmodernism*, 2nd ed., (Athens: University of Georgia Press, 1993).

Sassen, Saskia, *The Mobility of Labour and Capital: A Study of International Investment and Labour Flow* (Cambridge: Cambridge University Press, 1988).

Simons, Geoff, *Iraq: From Sumer to Post-Saddam* (New York: Palgrave Macmillan, 2004).

Singer, Peter, *The President of Good and Evil: The Ethics of George W. Bush* (New York: Dutton, 2004).

Singer, Peter W., *Corporate Warriors: The Rise of the Privatized Military Industry* (Ithaca, NY: Cornell University Press, 2003).

Slater, David, *Geopolitics and the Post-Colonial: Rethinking North-South Relations* (Malden, MA: Blackwell, 2004).

Smith, Neil, *The Endgame of Globalization* (New York: Routledge, 2005).

Soyinka, Wole, *Climate of Fear: The Quest for Dignity in a Dehumanized World* (New York: Random House, 2004).

Steger, Manfred B., *Globalism: Market Ideology Meets Terrorism*, 2nd ed., (Lanham, MD: Rowman & Littlefield, 2005).

Stephanson, Anders, *Manifest Destiny: American Expansion and the Empire of Right* (New York: Hill and Wang, 1995).

Tirman, John, "Introduction: The Movement of People and the Security of States," in *The Maze of Fear: Security and Migration after 9/11*, edited by John Tirman (New York: The New Press, 2004).

Tyner, James A., *Made in the Philippines: Gendered Discourses and the Making of Migrants* (London: Routledge, 2004).

————. *Iraq, Terror, and the Philippines' Will to War* (Boulder, CO: Rowman & Littlefield, 2005).

Wood, Ellen Meiksins, *Empire of Capital* (London: Verso, 2003).

Woodward, Peter N., *Oil and Labor in the Middle East: Saudi Arabia and the Oil Boom* (New York: Praeger, 1988).

Woodward, Rachel, "From Military Geography to Militarism's Geographies: Disciplinary Engagements with the Geographies of Militarism and Military Activities," *Progress in Human Geography* 29 (2005): 718-740.

Young, Robert J.C., *Post-Colonialism: A Very Short Introduction* (Oxford: Oxford University Press, 2003).

Zizek, Slavoj, *Iraq: The Borrowed Kettle* (New York: Verso, 2004).

Index